Quarks, Leptons and
the Big Bang
Second Edition

Quarks, Leptons and the Big Bang

Second Edition

Jonathan Allday

The King's School, Canterbury

Taylor & Francis
Taylor & Francis Group
New York   London

Published in 2002 by
Taylor & Francis Group
270 Madison Avenue
New York, NY 10016

Published in Great Britain by
Taylor & Francis Group
2 Park Square
Milton Park, Abingdon
Oxon OX14 4RN

International Standard Book Number-10: 0-7503-0806-0 (Softcover)
International Standard Book Number-13: 978-0-7503-0806-9 (Softcover)

Library of Congress Cataloging-in-Publication Data

Catalog record is available from the Library of Congress

Visit the Taylor & Francis Web site at
http://www.taylorandfrancis.com

Contents

Preface to the second edition

It is surely a truism that if you wrote a book twice, you would not do it the same way the second time. In my case, *Quarks, Leptons and the Big Bang* was hauled round several publishers under the guise of a textbook for schools in England. All the time I knew that I really wanted it to be a popular exposition of particle physics and the big bang, but did not think that publishers would take a risk on such a book from an unknown author. Well, they were not too keen on taking a risk with a textbook either. In the end I decided to send it to IOPP as a last try. Fortunately Jim Revill contacted me to say that he liked the book, but thought it should be more of a popular exposition than a textbook...

This goes some way to explaining what some have seen as slightly odd omissions from the material in this book—some mention of superstrings as one example. Such material was not needed in schools and so did not make it into the book. However, now that we are producing a second edition there is a chance to correct that and make it a little more like it was originally intended to be.

I am very pleased to say that the first edition has been well received. Reviewers have been kind, sales have been satisfying and there have been many emails from people saying how much they enjoyed the book. Sixth form students have written to say they like it, a University of the Third Age adopted it as a course book and several people have written to ask me further questions (which I tried to answer as best I could). It has been fun to have my students come up to me from time to time to say that they have found one of my books on the Amazon web site and (slightly surprised tone of voice) the reviewers seem to *like* it.

Well here goes with a second edition. As far as particle physics is concerned nothing has changed fundamentally since the first edition was published. I have taken the opportunity to add some material on field theory and to tweak the chapters on forces and quantum theory. The information on what is going on at CERN has been brought more up to date including some comment on the Higgs 'discovery' at CERN. There are major revisions to the cosmology sections that give more balance to the two aspects of the book. In the first edition cosmology was dealt with in two chapters; now it has grown to chapters 12, 13, 14 and 15. The new chapter 13 introduces general relativity in far more detail and bolsters the coverage of how it applies to cosmology. The evidence for dark matter has been pulled together into chapter 14 and brought more up to date by adding material on gravitational lensing. Inflation is dealt with in chapter 15. Experimental support for inflation has grown and there is now strong evidence to suggest that Einstein's cosmological constant is going to have to be dusted off. All this is covered in the final chapter of the book.

There are some quite exciting times ahead for cosmologists as the results of new experiments probing the background radiation start to come in over the next few years. Probably something really important will happen just after the book hits the shelves.

Then there will have to be a third edition...

Further thanks

Carlos S Frenk	(University of Durham) who spent some of his valuable time reading the cosmology sections of the first edition and then helping me bring them up to date.
Andrew Liddle	(University of Sussex) for help with certain aspects of inflationary theory.
Jim Revill	For continual support and encouragement at IOPP.
Simon Laurenson	Continuing the fine production work at IOPP.
Carolyn Allday	Top of the 'without whom' list.
Toby Allday	Another possible computer burner who held off.

Jonathan Allday
Jonathan.Allday@btinternet.com
Sunday, April 15, 2001

Preface to the first edition

It is difficult to know what to say in the preface to a book. Certainly it should describe what the book is about.

This is a book about particle physics (the strange world of objects and forces that exists at length scales much smaller than the size of an atom) and cosmology (the study of the origin of the universe). It is quite extraordinary that these two extremes of scale can be drawn together in one book. Yet the advances of the past couple of decades have shown that there is an intimate relationship between the world of the very large and the very small. The key moment that started the forging of this relationship was the discovery of the expansion of the universe in the 1920s. If the universe has been expanding since its creation (some 15 billion years ago) then at some time in the past the objects within it were very close together and interacting by the forces that particle physicists study. At one stage in its history the whole universe was the microscopic world. In this book I intend to take the reader on a detailed tour of the microscopic world and then through to the established ideas about the big bang creation of the universe and finally to some of the more recent refinements and problems that have arisen in cosmology. In order to do this we need to discuss the two most important fundamental theories that have been developed this century: relativity and quantum mechanics. The treatment is more technical than a popular book on the subject, but much less technical than a textbook.

Another thing that a preface should do is to explain what the reader is expected to know in advance of starting this book.

In this book I have assumed that the reader has some familiarity with energy, momentum and force at about the level expected of a modern GCSE candidate. I have also assumed a degree of familiarity with mathematics—again at about the modern GCSE level. However, readers who are put off by mathematics can always leave the boxed calculations for another time without disturbing the thread of the argument.

Finally, I guess that the preface should give some clue as to the spirit behind the book. In his book *The Tao of Physics* Fritjof Capra says that physics is a 'path with a heart'. By this he means that it is a way of thinking that can lead to some degree of enlightenment not just about the world in which we live, but also about us, the people who live in it. Physics is a human subject, despite the dry mathematics and formal presentation. It is full of life, human tragedy, exhilaration, wonder and very hard work. Yet by and large these are not words that most people would associate with physics after being exposed to it at school (aside from hard work that is). Increasingly physics is being marginalized as an interest at the same time as it is coming to grips with the most fundamental questions of existence. I hope that some impression of the life behind the subject comes through in this book.

Acknowledgments

I have many people to thank for their help and support during the writing of this book.

Liz Swinbank, Susan Oldcorn and Lewis Ryder for their sympathetic reading of the book, comments on it and encouragement that I was on the right lines.

Professors Brian Foster and Ian Aitchison for their incredibly detailed readings that found mistakes and vagaries in the original manuscript. Thanks to them it is a much better book. Of course any remaining mistakes can only be my responsibility.

Jim Revill, Al Troyano and the production team at Institute of Physics Publishing.

Many students of mine (too many to list) who have read parts of the book. Various Open University students who have been a source of inspiration over the years and a captive audience when ideas that ended up in this book have been put to the test at summer schools.

Graham Farmello, Gareth Jones, Paul Birchley, David Hartley and Becky Parker who worked with Liz and I to spice up A level physics by putting particle physics in.

Finally thanks to family and friends.

Carolyn, Benjamin and Joshua who have been incredibly patient with me and never threatened to set fire to the computer.

My parents Joan and Frank who knew that this was something that I really wanted to do.

John and Margaret Gearey for welcoming me in.

Robert James, a very close friend for a very long time.

Richard Houlbrook, you see I said that I would not forget you.

Jonathan Allday
November 1997

Prelude

Setting the scene

What is particle physics?

Particle physics attempts to answer some of the most basic questions about the universe:

- are there a small number of different types of objects from which the universe is made?
- do these objects interact with each other and, if so, are there some simple rules that explain what will happen?
- how can we study the creation of the universe in a laboratory?

The topics that particle physicists study from one day to the next have changed as the subject has progressed, but behind this progression the final goal has remained the same—to try to understand how the universe came into being.

Particle physics tries to answer questions about the origin of our universe by studying the objects that are found in it and the ways in which they interact. This is like someone trying to learn how to play chess by studying the shapes of the pieces and the ways in which they move across the board.

Perhaps you think that this is a strange way to try to find out about the origin of the universe. Unfortunately, there is no other way. There are instruction manuals to help you learn how to play chess; there are no instruction manuals supplied with the universe. Despite this handicap an impressive amount has been understood by following this method.

1

Some people argue that particle physics is fundamental to all the sciences as it strips away the layers of structure that we see in the world and plunges down to the smallest components of matter. This study applies equally to the matter that we see on the earth and that which is in the stars and galaxies that fill the whole universe. The particle physicist assumes that all matter in the universe is fundamentally the same and that it all had a common origin in the big bang that created our universe. (This is a reasonable assumption as we have no evidence to suggest that any region of the universe is made of a different form of matter. Indeed we have positive evidence to suggest the opposite.)

The currently accepted scientific theory is that our universe came into being some fifteen billion years ago in a gigantic explosion. Since then it has been continually growing and cooling down. The matter created in this explosion was subjected to unimaginable temperatures and pressures. As a result of these extreme conditions, reactions took place that were crucial in determining how the universe would turn out. The structure of the universe that we see now was determined just after its creation.

If this is so, then the way that matter is structured now must reflect this common creation. Hence by building enormous and expensive accelerating machines and using them to smash particles together at very high energies, particle physicists can force the basic constituents of matter into situations that were common in the creation of the universe— they produce miniature big bangs. Hardly surprisingly, matter can behave in very strange ways under these circumstances.

Of course, this programme was not worked out in advance. Particle physics was being studied before the big bang theory became generally accepted. However, it did not take long before particle physicists realized that the reactions they were seeing in their accelerators must have been quite common in the early universe. Such experiments are now providing useful information for physicists working on theories of how the universe was created.

In the past twenty years this merging of subjects has helped some huge leaps of understanding to take place. We believe that we have an accurate understanding of the evolution of the universe from the first 10^{-5} seconds onwards (and a pretty good idea of what happened even earlier). By the time you have finished this book, you will have met many of the basic ideas involved.

Why study particle physics?

All of us, at some time, have paused to wonder at our existence. As children we asked our parents embarrassing questions about where we came from (and, in retrospect, probably received some embarrassing answers). In later years we may ask this question in a more mature form, either in accepting or rejecting some form of religion. Scientists that dedicate themselves to pure research have never stopped asking this question.

It is easy to conclude that society does not value such people. Locking oneself away in an academic environment 'not connected with the real world' is generally regarded as a (poorly paid) eccentricity. This is very ironic. Scientists are engaged in studying a world far more real than the abstract shuffling of money on the financial markets. Unfortunately, the creation of wealth and the creation of knowledge do not rank equally in the minds of most people.

Against this background of poor financial and social status it is a wonder that anyone chooses to follow the pure sciences; their motivation must be quite strong. In fact, the basic motivation is remarkably simple.

Everyone has, at some time, experienced the inner glow that comes from solving a puzzle. This can take many forms, such as maintaining a car, producing a difficult recipe, solving a jigsaw puzzle, etc. Scientists are people for whom this feeling is highly magnified. Partly this is because they are up against the ultimate puzzle. As a practising and unrepentant physicist I can testify to the feeling that comes from prising open the door of nature by even a small crack and understanding something new for the first time. When such an understanding is achieved the feeling is one of personal satisfaction, but also an admiration for the puzzle itself. Few of us are privileged enough to get a glimpse through a half-open door, like an Einstein or a Hawking, but we can all look over their shoulders. The works of the truly great scientists are part of our culture and should be treated like any great artistic creation. Such work demands the support of society.

Unfortunately, the appreciation of such work often requires a high degree of technical understanding. This is why science is not valued as much as it might be. The results of scientific experiments are often felt to be beyond the understanding, and hence the interest, of ordinary people. Scientists are to blame. When Archimedes jumped out of his bath and ran through the streets shouting 'Eureka!' he did not stop to explain his actions to the passers by. Little has changed in this respect over the intervening centuries. We occasionally glimpse a scientist running past shouting about some discovery, but are unable to piece anything together from the fragments that we hear. Few scientists are any good at telling stories.

The greatest story that can be told is the story of creation. In the past few decades we have been given an outline of the plot, and perhaps a glimpse of the last page. As in all mystery stories the answer seems so obvious and simple, it is a wonder that we did not think of it earlier. This is a story so profound and wonderful that it must grab the attention of anyone prepared to give it a moment's time.

Once it has grabbed you, questions as to why we should study such things become irrelevant—*it is obvious that we must.*

Chapter 1

The standard model

This chapter is a brief summary of the theories discussed in the rest of this book. The standard model of particle physics— the current state of knowledge about the structure of matter— is described and an introduction provided to the 'big bang' theory of how the universe was created. We shall spend the rest of the book exploring in detail the ideas presented in this chapter.

1.1 The fundamental particles of matter

It is remarkable that a list of the fundamental constituents of matter easily fits on a single piece of paper. It is as if all the recipes of all the chefs that have been and will be could be reduced to combinations of twelve simple ingredients.

The twelve particles from which all forms of matter are made are listed in table 1.1. Twelve particles, that is all that there is to the world of matter.

The twelve particles are divided into two distinct groups called the *quarks* and the *leptons* (at this stage don't worry about where the names come from). Quarks and leptons are distinguished by the different ways in which they react to the fundamental forces.

There are six quarks and six leptons. The six quarks are called up, down, strange, charm, bottom and top[1] (in order of mass). The six leptons are

Table 1.1. The fundamental particles of matter.

Quarks		Leptons	
up	(u)	electron	(e^-)
down	(d)	electron-neutrino	(ν_e)
strange	(s)	muon	(μ^-)
charm	(c)	muon-neutrino	(ν_μ)
bottom	(b)	tau	(τ^-)
top	(t)	tau-neutrino	(ν_τ)

the electron, the electron-neutrino, the muon, muon-neutrino, tau and tau-neutrino. As their names suggest, their properties are linked.

Already in this table there is one familiar thing and one surprise.

The familiar thing is the electron, which is one of the constituents of the atom and the particle that is responsible for the electric current in wires. Electrons are fundamental particles, which means that they are not composed of any smaller particles—they do not have any pieces inside them. All twelve particles in table 1.1 are thought to be fundamental—they are all distinct and there are no pieces within them.

The surprise is that the proton and the neutron are not mentioned in the table. All matter is composed of atoms of which there are 92 naturally occurring types. Every atom is constructed from electrons which orbit round a small, heavy, positively charged nucleus. In turn the nucleus is composed of protons, which have a positive charge, and neutrons, which are uncharged. As the size of the charge on the proton is the same as that on the electron (but opposite in sign), a neutral atom will contain the same number of protons in its nucleus as it has electrons in its orbit. The numbers of neutrons that go with the protons can vary by a little, giving the different isotopes of the atom.

However, the story does not stop at this point. Just as we once believed that the atom was fundamental and then discovered that it is composed of protons, neutrons and electrons, we now know that the protons and neutrons are not fundamental either (but the electron is, remember). Protons and neutrons are composed of quarks.

Specifically, the proton is composed of two up quarks and one down quark. The neutron is composed of two down quarks and one up quark. Symbolically we can write this in the following way:

$$p \equiv uud$$
$$n \equiv udd.$$

As the proton carries an electrical charge, at least some of the quarks must also be charged. However, similar quarks exist inside the neutron, which is uncharged. Consequently the charges of the quarks must add up in the combination that composes the proton but cancel out in the combination that composes the neutron. Calling the charge on an up quark Q_u and the charge on a down quark Q_d, we have:

$$p \text{ (uud) charge} = Q_u + Q_u + Q_d = 1$$
$$n \text{ (udd) charge} = Q_u + Q_d + Q_d = 0.$$

Notice that in these relationships we are using a convention that sets the charge on the proton equal to $+1$. In standard units this charge would be approximately 1.6×10^{-19} coulombs. Particle physicists normally use this abbreviated unit and understand that they are working in multiples of the proton charge (the proton charge is often written as $+e$).

These two equations are simple to solve, producing:

$$Q_u = \text{charge on the up quark} = +\tfrac{2}{3}$$
$$Q_d = \text{charge on the down quark} = -\tfrac{1}{3}.$$

Until the discovery of quarks, physicists thought that electrical charge could only be found in multiples of the proton charge. The standard model suggests that there are three basic quantities of charge: $+2/3$, $-1/3$ and -1.[2]

The other quarks also have charges of $+2/3$ or $-1/3$. Table 1.2 shows the standard way in which the quarks are grouped into families. All the quarks in the top row have charge $+2/3$, and all those in the bottom row have charge $-1/3$. Each column is referred to as a *generation*. The up and down quarks are in the first generation; the top and bottom quarks belong to the third generation.

Table 1.2. The grouping of quarks into generations (NB: the letters in brackets are the standard abbreviations for the names of the quarks).

	1st generation	2nd generation	3rd generation
$+2/3$	up (u)	charm (c)	top (t)
$-1/3$	down (d)	strange (s)	bottom (b)

This grouping of quarks into generations roughly follows the order in which they were discovered, but it has more to do with the way in which the quarks respond to the fundamental forces.

All the matter that we see in the universe is composed of atoms—hence protons and neutrons. Therefore the most commonly found quarks in the universe are the up and down quarks. The others are rather more massive (the mass of the quarks increases as you move from generation 1 to generation 2 and to generation 3) and very much rarer. The other four quarks were discovered by physicists conducting experiments in which particles were made to collide at very high velocities, producing enough energy to make the heavier quarks.

In the modern universe heavy quarks are quite scarce outside the laboratory. However, earlier in the evolution of the universe matter was far more energetic and so these heavier quarks were much more common and had significant roles to play in the reactions that took place. This is one of the reasons why particle physicists say that their experiments allow them to look back into the history of the universe.

We should now consider the leptons. One of the leptons is a familiar object—the electron. This helps in our study of leptons, as the properties of the electron are mirrored in the muon and the tau. Indeed, when the muon was first discovered a famous particle physicist was heard to remark 'who ordered that?'. There is very little, besides mass, that distinguishes the electron from the muon and the tau. They all have the same electrical charge and respond to the fundamental forces in the same way. The only obvious difference is that the muon and the tau are allowed to decay into other particles. The electron is a stable object.

Aside from making the number of leptons equal to the number of quarks, there seems to be no reason why the heavier leptons should exist. The heavy quarks can be found in some exotic forms of matter and detailed theory requires that they exist—but there is no such apparent constraint on the leptons. The heavy quarks can be found in some exotic forms of matter and detailed theory requires that they exist—but there is no such apparent constraint on the leptons. It is a matter of satisfaction to physicists that there are equal numbers of quarks and leptons, but there is no clear idea at this stage why this should be so. This 'coincidence' has suggested many areas of research that are being explored today.

The other three leptons are all called neutrinos as they are electrically neutral. This is not the same as saying, for example, that the neutron has a zero charge. A neutron is made up of three quarks. Each of these quarks carries an electrical charge. When a neutron is observed from a distance, the electromagnetic effects of the quark charges balance out making the neutron look like a neutral object. Experiments that probe inside the neutron can resolve the presence of charged objects within it. Neutrinos, on the other hand, are fundamental particles. They have no components inside them—they are *genuinely* neutral. To distinguish such particles from ones whose component charges cancel, we shall say that the neutrinos (and particles like them) are *neutral*, and that neutrons (and particles like them) have *zero charge*.

Neutrinos have extremely small masses, even on the atomic scale. Experiments with the electron-neutrino suggest that its mass is less than one ten-thousandth of that of the electron. Many particle physicists believe that the neutrinos have no mass at all. This makes them the most ghost-like objects in the universe. Many people are struck by the fact that neutrinos have no charge or mass. This seems to deny them any physical existence at all! However, neutrinos do have energy and this energy gives them reality.

The names chosen for the three neutrinos suggest that they are linked in some way to the charged leptons. The link is formed by the ways in which the leptons respond to one of the fundamental forces. This allows us to group the leptons into generations as we did with the quarks. Table 1.3 shows the lepton generations.

Table 1.3. The grouping of leptons into generations (NB: the symbols in brackets are the standard abbreviations for the names of the leptons).

	1st generation	2nd generation	3rd generation
-1	electron (e^-)	muon (μ^-)	tau (τ^-)
0	electron-neutrino (ν_e)	muon-neutrino (ν_μ)	tau-neutrino (ν_τ)

The masses of the leptons increase as we move up the generations (at least this is true of the top row; as noted above, it is still an open question whether the neutrinos have any mass at all).

At this stage we need to consider the forces by which all these fundamental particles interact. This will help to explain some of the reasons for grouping them in the generations (which, incidentally, will make the groups much easier to remember).

1.2 The four fundamental forces

A fundamental force cannot be explained as arising from the action of a more basic type of force. There are many forces in physics that occur in different situations. For example:

- gravity;
- friction;
- tension;
- electromagnetic[3];
- van der Waals.

Only two of the forces mentioned in this list (gravity and electromagnetic) are regarded as fundamental forces. The rest arise due to more fundamental forces.

For example, friction takes place when one object tries to slide over the surface of another. The theory of how frictional forces arise is very complex, but in essence they are due to the electromagnetic forces between the atoms of one object and those of another. Without electromagnetism there would be no friction.

Similarly, the tensional forces that arise in stretched wires are due to electromagnetic attractions between atoms in the structure of the wire. Without electromagnetism there would be no tension. Van der Waals forces are the complex forces that exist between atoms or molecules. It is the van der Waals attraction between atoms or molecules in a gas that allow the gas to be liquefied under the right conditions of temperature and pressure. These forces arise as a combination of the electromagnetic repulsion between the electrons of one atom and the electrons of another and the attraction between the electrons of one atom and the nucleus of another. Again the theory is quite complex, but the forces arise out of the electromagnetic force in a complex situation. Without the electromagnetic force there would be no van der Waals forces.

These examples illustrate the difference between a force and a fundamental force. Just as a fundamental particle is one that is not composed of any pieces, a fundamental force is one that does not arise out of a more basic force.

Particle physicists hope that one day they will be able to explain all forces out of the action of just one fundamental force. In chapter 10 we will see how far this aim has been achieved.

The standard model recognizes four forces as being sufficiently distinct and basic to be called fundamental forces:

- gravity;
- electromagnetic;
- the weak force;
- the strong force.

Our experiments indicate that these forces act in very different ways from each other at the energies that we are currently able to achieve. However, there is some theoretical evidence that in the early history of the universe particle reactions took place at such high energies that the forces started to act in very similar ways. Physicists regard this as an indication that there is one force, more fundamental than the four listed above, that will eventually be seen as the single force of nature.

Gravity and electromagnetism will already be familiar to you. The weak and strong forces may well be new. In the history of physics they have only recently been discovered. This is because these forces have a definite *range*—they only act over distances smaller than a set limit. For distances greater than this limit the forces become so small as to be undetectable.

The range of the strong force is 10^{-15} m and that of the weak force 10^{-17} m. A typical atom is about 10^{-10} m in diameter, so these forces have ranges smaller than atomic sizes. A proton on one side of a nucleus would be too far away from a proton on the other side to interact through the action of the weak force! Only in the last 60 years have we been able to conduct experiments over such short distances and observe the physics of these forces.

1.2.1 The strong force

Two protons placed 1 m apart from each other would electromagnetically repel with a force some 10^{42} times greater than the gravitational attraction between them. Over a similar distance the strong force would be zero. If, however, the distance were reduced to a typical nuclear diameter, then the strong force would be at least as big as the electromagnetic. It is the strong force attraction that enables a nucleus, which packs protons into a small volume, to resist being blown apart by electrostatic repulsion.

The strong force only acts between quarks. The leptons do not experience the strong force at all. They are effectively blind to it (similarly a neutral object does not experience the electromagnetic force). This is the reason for the division of the material particles into the quarks and leptons.

➡   Quarks feel the strong force, leptons do not.
➡   Both quarks and leptons feel the other three forces.

This incredibly strong force acting between the quarks holds them together to form objects (particles) such as the proton and the neutron. If the leptons could feel the strong force, then they would also bind together into particles. This is the major difference between the properties of the quarks and leptons.

➡   The leptons do not bind together to form particles.
➡   The strong force between quarks means that they can *only* bind together to form particles.

Current theories of the strong force suggest that it is impossible to have a single quark isolated without any other quarks. All the quarks in the universe at the moment are bound up with others into particles. When we create new quarks in our experiments, they rapidly combine with others. This happens so quickly that it is impossible to ever see one on its own.

The impossibility of finding a quark on its own makes them very difficult objects to study. Some of the important experimental techniques used are discussed in chapter 8.

1.2.2   The weak force

The weak force is the most difficult of the fundamental forces to describe. This is because it is the one that least fits into our typical imagination of what a force should do. It is possible to imagine the strong force as being an attractive force between quarks, but the categories 'attractive' and 'repulsive' do not really fit the weak force. This is because it changes particles from one type to another.

The weak force is the reason for the generation structure of the quarks and leptons. The weak force is felt by both quarks and leptons. In this respect it is the same as the electromagnetic and gravitational forces— the strong force is the only one of the fundamental forces that is only felt by one class of material particle.

If two leptons come within range of the weak force, then it is possible for them to be changed into other leptons, as illustrated in figure 1.1.

Figure 1.1 is deliberately suggestive of the way in which the weak force operates. At each of the black blobs a particle has been changed from one type into another. In the general theory of forces (discussed in chapter 10) the 'blobs' are called *vertices*. The weak force can change a particle from one type into another at such a vertex. However, it is only possible for the weak force to change leptons *within the same generation* into each other. The electron can be turned into an electron-neutrino, and vice versa, but the electron cannot be turned into the muon-neutrino (or a muon for that matter). This is why we divide the leptons into generations. The weak force can act within the lepton generations, but not between them.

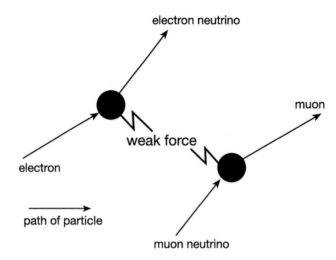

Figure 1.1. A representation of the action of the weak force.

There is a slight complication when it comes to the quarks. Again the weak force can turn one quark into another and again the force acts within the generations of quarks. However, it is not true to say that the force cannot act across generations. It can, but with a much reduced effect. Figure 1.2 illustrates this.

The generations are not as strictly divided in the case of quarks as in the case of leptons—there is less of a generation gap between quarks.

This concludes a very brief summary of the main features of the four fundamental forces. They will be one of the key elements in our story and we will return to them in increasing detail as we progress.

1.3 The big bang

Our planet is a member of a group of nine planets that are in orbit round a star that we refer to as *the sun*. This family of star and planets we call our *solar system*. The sun belongs to a *galaxy* of stars (in fact it sits on the edge of the galaxy—out in the suburbs) many of which probably have solar systems of their own. In total there are something like 10^{11} stars in our galaxy. We call this galaxy the *Milky Way*.

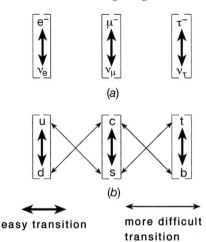

Figure 1.2. (*a*) The effect of the weak force on the leptons. (*b*) The effect of the weak force on the quarks. (NB: transformations such as u → b are also possible.)

The Milky Way is just one of many galaxies. It is part of the 'local group', a collection of galaxies held together by gravity. There are 18 galaxies in the local group. Astronomers have identified other clusters of galaxies, some of which contain as many as 800 separate galaxies loosely held together by gravity.

It is estimated that there are 3×10^9 galaxies observable from earth. Many of these are much bigger than our Milky Way. Multiplied together this means that there are $\sim 10^{20}$ stars in the universe (if our galaxy is, as it appears to be, of a typical size). There are more stars in the universe than there are grains of sand on a strip of beach.

The whole collection forms what astronomers call the *observable universe*. We are limited in what we can see by two issues. Firstly, the further away a galaxy is the fainter the light from it and so it requires a large telescope to see it. The most distant galaxies have been resolved by the Hubble Space Telescope in Earth orbit, which has seen galaxies that are so far away light has been travelling nearly 14 billion years to get to us. The second limit on our view is the time factor. Our best current estimates put the universe at something like 15 billion years old—so any

galaxy more than 15 billion light years[4] away cannot be seen as the light
has not had time to reach us yet!

Undoubtedly the actual universe is bigger than the part that we can
currently observe. Just how much bigger is only recently becoming
apparent. If our latest ideas are correct, then the visible universe is
just a tiny speck in an inconceivably vaster universe. (For the sake of
brevity from now on I shall use the term 'universe' in this book rather
than 'visible universe'.)

Cosmology is an unusual branch of physics[5]. Cosmologists study the
possible ways in which the universe could have come about, how it
evolved and the various possible ways in which it will end (including that
it won't!). There is scarcely any other subject which deals with themes
of such grandeur. It is an extraordinary testimony to the ambition of
science that this subject has become part of mainstream physics.

With contributions from the theories of astronomy and particle physics,
cosmologists are now confident that they have a 'standard model' of how
the universe began. This model is called the *big bang*.

Imagine a time some fifteen billion years ago. All the matter in the
universe exists in a volume very much smaller than it does now—smaller,
even, than the volume of the earth. There are no galaxies, indeed no
matter as we would recognize it from our day-to-day experience at al.

The temperature of the whole universe at this time is incredible, greater
than 10^{33} K. The only material objects are elementary particles reacting
with each other more often and at higher energies than we have ever been
able to reproduce in our experiments (at such temperatures the kinetic
energy of a single particle is greater than that of a jet plane).

The universe is expanding—not just a gradual steady expansion,
an explosive expansion that causes the universe to double in size
every 10^{-34} seconds. This period lasts for a fleeting instant (about
10^{-35} seconds), but in this time the universe has grown by a factor of
10^{50}. At the end of this extraordinary period of inflation the matter
that was originally in the universe has been spread over a vast scale.
The universe is empty of matter, but full of energy. As we will see in
chapter 15 this energy is unstable and it rapidly collapses into matter of

a more ordinary form re-filling the universe with elementary particles at a tremendous temperature.

From now on the universe continues to expand, but at a more leisurely pace compared to the extraordinary inflation that happened earlier. The universe cools as it expands. Each particle loses energy as the gravity of the rest of the universe pulls it back. Eventually the matter cools enough for the particles to combine in ways that we shall discuss as this book progresses—matter as we know it is formed.

However, the inflationary period has left its imprint. Some of the matter formed by the collapse of the inflationary energy is a mysterious form of *dark matter* that we cannot see with telescopes. Gravity has already started to gather this dark matter into clumps of enormous mass seeded by tiny variations in the amount of inflationary energy from place to place. Ordinary matter is now cool enough for gravity to get a grip on it as well and this starts to fall into the clumps of dark matter. As the matter gathers together inside vast clouds of dark matter, processes start that lead to the evolution of galaxies and the stars within them. Eventually man evolves on his little semi-detached planet.

What caused the big bang to happen? We do not know. Our theories of how matter should behave do not work at the temperatures and pressures that existed just after the big bang. At such small volumes all the particles in the universe were so close to each other that gravity plays a major role in how they would react. As yet we are not totally sure of how to put gravity into our theories of particle physics. There is a new theory that seems to do the job (superstring theory) and it is starting to influence some cosmological thinking, but the ideas are not yet fully worked out.

It is difficult to capture the flavour of the times for those who are not directly involved. Undoubtedly the last 30 years have seen a tremendous advance in our understanding of the universe. The very fact that we can now theorize about how the universe came about is a remarkable step forward in its own right. The key to this has been the unexpected links that have formed between the physics of elementary particles and the physics of the early universe. In retrospect it seems inevitable. We are experimenting with reacting particles together in our laboratories and trying to re-create conditions that existed quite naturally in the early universe. However, it also works the other way. We can observe the current state of the universe and theorize about how the structures we see came about. This leads inevitably to the state of the early universe and so places constraints on how the particles must have reacted then.

A great deal has been done, but a lot is left to do. The outlines of the process of galaxy formation have been worked out, but some details have to be filled in. The dark matter referred to earlier is definitely present in the universe, but we do not know what it is—it needs to be 'discovered' in experiments on Earth. Finally the very latest results (the past couple of years) suggest that there may be yet another form of energy that is present in the universe that is exerting a gravitational repulsion on the galaxies causing them to fly apart faster and faster. As experimental results come in over the next decade the existence of this dark energy will be confirmed (or refuted) and the theorists will get to work.

Finally some physicists are beginning to seriously speculate about what the universe may have been like before the big bang...

1.4 Summary of chapter 1

- There are two types of fundamental material particle: quarks and leptons;
- fundamental particles do not contain any other objects within them;
- there are four fundamental forces: strong, weak, electromagnetic and gravity;
- fundamental forces are not the result of simpler forces acting in complicated circumstances;
- there are six different quarks and six different leptons;
- the quarks can be divided into three pairings called generations;
- the leptons can also be divided into three generations;
- the quarks feel the strong force, the leptons do not;
- both quarks and leptons feel the other three forces;
- the strong force binds quarks together into particles;
- the weak force can turn one fundamental particle into another—but in the case of leptons it can act only within generations, whereas with quarks it predominantly acts within generations;
- the weak force cannot turn quarks into leptons or vice versa;
- the universe was, we believe, created some 15 billion years ago;

- the event of creation was a gigantic 'explosion'—the big bang—in which all the elementary particles were produced;
- as a result of this 'explosion' the bulk of the matter in the universe is flying apart, even today;
- the laws of particle physics determine the early history of the universe.

Notes

[1] Physicists are not very good at naming things. Over the past few decades there seems to have been an informal competition to see who can come up with the silliest name for a property or a particle. This is all harmless fun. However, it does create the illusion that physicists do not take their jobs seriously. Being semi-actively involved myself, I am delighted that physicists are able to express their humour and pleasure in the subject in this way—it has been stuffy for far too long! However, I do see that it can cause problems for others. Just remember that the actual names are not important—it is what the particles *do* that counts!

Murray Gell-Mann has been at the forefront of the 'odd names' movement for several years. He has argued that during the period when physicists tried to name things in a meaningful way they invariably got it wrong. For example atoms, so named because they were indivisible, were eventually split. His use of the name 'quark' was a deliberate attempt to produce a name that did not mean anything, and so could not possibly be wrong in the future! The word is actually taken from a quotation from James Joyce's *Finnegan's Wake*: 'Three quarks for Muster Mark'.

[2] It is a very striking fact that the total charge of 2u quarks and 1d quark (the proton) should be exactly the same size as the charge on the electron. This is very suggestive of some link between the quarks and leptons. There are some theories that make a point of this link, but as yet there is no experimental evidence to support them.

[3] The electromagnetic force is the name for the combined forces of electrostatics and magnetism. The complete theory of this force was developed by Maxwell in 1864. Maxwell's theory drew all the separate areas of electrostatics, magnetism and electromagnetic induction together, so we now tend to use the terms electromagnetism or electromagnetic force to refer to all of these effects.

[4] The light year is a common distance unit used by astronomers. Some people are confused and think that a light year is a measurement of time, it is not—light

years measure *distance*. One light year is the distance travelled by a beam of light in one year. As the speed of light is three hundred million metres per second, a simple calculation tells us that a light year is a distance of 9.44×10^{15} m. On this scale the universe is thought to be roughly 2.0×10^{10} light years in diameter.

[5] It is very difficult to carry out experiments in cosmology! I have seen a spoof practical examination paper that included the question: 'given a large energy source and a substantial volume of free space, prepare a system in which life will evolve within 15 billion years'.

Chapter 2

Aspects of the theory of
relativity

In this chapter we shall develop the ideas of special relativity
that are of most importance to particle physics. The aim will be
to understand these ideas and their implications, not to produce
a technical derivation of the results. We will not be following
the historically correct route. Instead we will consider two
experiments that took place after Einstein published his theory.

Unfortunately there is not enough space in a book specifically about
particle physics to dwell on the strange and profound features of
relativity. Instead we shall have to concentrate on those aspects of the
theory that are specifically relevant to us. The next two chapters are
more mathematical than any of the others in the book. Those readers
whose taste does not run to algebraic calculation can simply miss out the
calculations, at least on first reading.

2.1 Momentum

The first experiment that we are going to consider was designed to
investigate how the momentum of an electron varied with its velocity.
We will use the results to explain one of the cornerstones of relativity.

Momentum as defined in Newtonian mechanics is the product of an object's mass and its velocity:

$$\text{momentum} = \text{mass} \times \text{velocity}$$

$$p = mv. \tag{2.1}$$

In a school physics experiment, momentum would be obtained by measuring the mass of an object when stationary, measuring its velocity while moving and multiplying the two results together. Experiments carried out in this way invariably demonstrate Newton's second law of motion in the form:

$$\text{applied force} = \text{rate of change of momentum}$$

$$f = \frac{\Delta(mv)}{\Delta t}. \tag{2.2}$$

However, there is a way to measure the momentum of a charged particle *directly*, rather than via measuring its mass and velocity separately. The technique relies on the force exerted on a charged particle moving through a magnetic field.

A moving charged particle passing through a magnetic field experiences a force determined by the charge of the particle, the speed with which it is moving and the size of the magnetic field[1]. Specifically:

$$F = Bqv \tag{2.3}$$

where B = size of magnetic field, q = charge of particle, v = speed of particle. The direction of this force is given by Fleming's left-hand rule (figure 2.1). The rule is defined for a positively charged particle. If you are considering a negative particle, then the direction of motion must be reversed—i.e. a negative particle moving to the right is equivalent to a positive particle moving to the left.

If the charged particle enters the magnetic field travelling at right angles to the magnetic field lines, the force will always be at right angles to both the field and the direction of motion (figure 2.2). Any force that is always at 90° to the direction of motion is bound to move an object in a circular path. The charged particle will be deflected as it passes through the magnetic field and will travel along the arc of a circle.

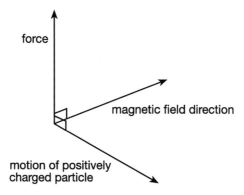

Figure 2.1. The left-hand rule showing the direction of force acting on a moving charge.

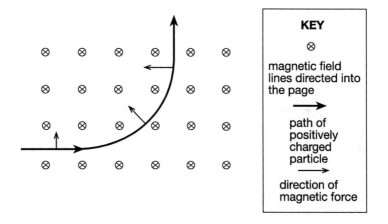

Figure 2.2. The motion of a charged particle in a magnetic field.

The radius of this circular path is a direct measure of the momentum of the particle. This can be shown in the context of Newtonian mechanics to be:

$$r = \frac{p}{Bq}$$

where r = radius of the path, p = momentum of the particle, B = magnetic field strength. A proof of this formula follows for those of a mathematical inclination.

To move an object of mass m on a circular path of radius r at a speed v, a force must be provided of size:

$$F = \frac{mv^2}{r}.$$

In this case, the force is the magnetic force exerted on the charged particle

$$F = Bqv$$

therefore

$$Bqv = \frac{mv^2}{r}$$

making r the subject:

$$r = \frac{mv}{Bq}$$

or

$$r = \frac{p}{Bq}$$

where p is the momentum of the particle. Equally:

$$p = Bqr.$$

This result shows that measuring the radius of the curve on which the particle travels, r, will provide a direct measure of the momentum, p, provided the size of the magnetic field and the charge of the particle are known.

This technique is used in particle physics experiments to measure the momentum of particles. Modern technology allows computers to reproduce the paths followed by particles in magnetic fields. It is then a simple matter for the software to calculate the momentum. When such experiments were first done (1909), much cruder techniques had to be used. The tracks left by the particles as they passed through photographic films were measured by hand to find the radius of the tracks. A student's thesis could consist of the analysis of a few such photographs.

The basis of the experiment is therefore quite simple: accelerate electrons to a known speed and let them pass through a magnetic field to measure their momentum. Electrons were used as they are lightweight particles with an accurately measured electrical charge.

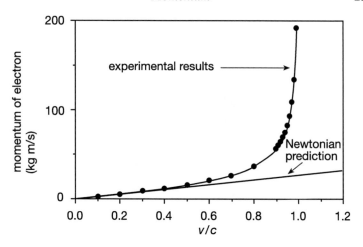

Figure 2.3. Momentum of an electron as a function of speed (NB: the horizontal scale is in fractions of the speed of light, *c*).

Figure 2.3 shows the data produced by a series of experiments carried out between 1909 and 1915[2]. The results are startling.

The simple Newtonian prediction shown on the graph is quite evidently wrong. For small velocities the experiment agrees well with the Newtonian formula for momentum, but as the velocity starts to get larger the difference becomes dramatic. The most obvious disagreement lies close to the magic number 3×10^8 m s^{-1}, the speed of light.

There are two possibilities here:

(1) the relationship between the momentum of a charged particle and the radius of curvature breaks down at speeds close to that of light;
(2) the Newtonian formula for momentum is wrong.

Any professional physicist would initially suspect the first possibility above the second. Although this is a reasonable suspicion, it is wrong. It is the second possibility that is true.

The formula that correctly follows the data is:

$$p = \frac{mv}{\sqrt{1 - v^2/c^2}} \tag{2.4}$$

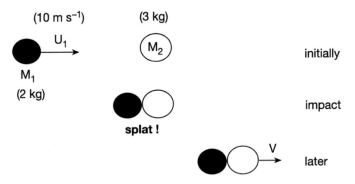

Figure 2.4. Collision between two particles that stick together.

where **p** is the relativistic momentum. In this equation c^2 is the velocity of light squared, or approximately 9.0×10^{16} m^2 s^{-2}, a huge number! Although this is not a particularly elegant formula it does follow the data very accurately. This is the momentum formula proposed by the theory of relativity.

This is a radical discovery. Newtonian momentum is *defined* by the formula mv, so it is difficult to see how it can be *wrong*. Physicists adopted the formula $p = mv$ because it seemed to be a good description of nature; it helped us to calculate what might happen in certain circumstances.

One of the primary reasons why we consider momentum to be useful is because it is a *conserved quantity*. Such quantities are very important in physics as they allow us to compare situations before and after an interaction, without having to deal directly with the details of the interaction.

Consider a very simple case. Figure 2.4 shows a moving particle colliding and sticking to another particle that was initially at rest.

The way in which particles stick together can be very complex and to study it in depth would require a sophisticated understanding of the molecular structure of the materials involved. However, if we simply want to calculate the speed at which the combined particles are moving after they stick together, then we can use the fact that momentum is

conserved. The total momentum before the collision must be the same
as that after the collision.

Using the Newtonian momentum formula, we can calculate the final
speed of the combined particle in the following way:

$$\text{initial momentum} = M_1 U_1$$
$$= 2 \text{ kg} \times 10 \text{ m s}^{-1}$$
$$= 20 \text{ kg m s}^{-1}$$
$$\text{final momentum} = (M_1 + M_2)V$$
$$= 5 \text{ kg} \times V.$$

Momentum is conserved, therefore

$$\text{initial momentum} = \text{final momentum}.$$

Therefore

$$5 \text{ kg} \times V = 20 \text{ kg m s}^{-1}$$
$$V = \frac{20 \text{ kg m s}^{-1}}{5 \text{ kg}}$$
$$= 4 \text{ m s}^{-1}.$$

It is easy to imagine doing an experiment that would confirm the results
of this calculation. On the basis of such experiments, we accept as a law
of nature that momentum *as defined by* $p = mv$ is a conserved quantity.
If mv were not conserved, then we would not bother with it—it would
not be a useful number to calculate.

However, modern experiments show that it does not give the correct
answers if the particles are moving quickly. We have seen how the
magnetic bending starts to go wrong, but this can also be confirmed by
simple collision experiments like that shown in figure 2.4. They also
go wrong when the speeds become significant fractions of the speed of
light.

For example, if the initially moving particle has a velocity of half the
speed of light ($c/2$), then after it has struck the stationary particle the
combined object should have a speed of $c/5$ (0.2c). However, if we carry
out this experiment, then we find it to be 0.217c. The Newtonian formula

is *wrong*—it predicted the incorrect velocity. The correct velocity can be obtained by using the relativistic momentum instead of the Newtonian one.

Experiments are telling us that the quantity mv does not always correspond to anything useful—it is not always conserved. When the velocity of a particle is small compared with light (and the speed of light is so huge, so that covers most of our experience!), the Newtonian formula *is a good approximation to the correct one*[3]. This is why we continue to use it and to teach it. The correct formula is rather tiresome to use, so when the approximation $p \approx mv$ gives a good enough result it seems silly not to use it. The mistake is to act as if it will *always* give the correct answer.

There are other quantities that are conserved (energy for one) in particle physics reactions. Before the standard model was developed much of particle physics theory was centred round the study of conserved quantities.

2.2   Kinetic energy

Our second experiment considers the kinetic energy of a particle. Conventional theory tells us that when an object is moved through a certain distance by the application of a force, energy is transferred to the object. This is known as 'doing work':

$$\text{work done} = \text{force applied} \times \text{distance moved.}$$

i.e.

$$W = F \times X. \tag{2.5}$$

The energy transferred in this manner increases the kinetic energy of the object:

$$\text{work done} = \text{change in kinetic energy.} \tag{2.6}$$

In Newtonian physics it is a simple matter to calculate the change in the kinetic energy, T, from Newton's second law of motion and the momentum, p. The result of such a calculation is[4]:

$$T = \tfrac{1}{2}mv^2 = \frac{p^2}{2m}. \tag{2.7}$$

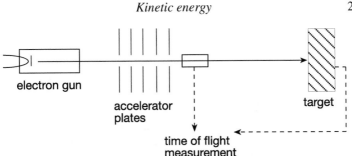

Figure 2.5. An energy transfer experiment.

We can check this formula by accelerating electrons in an electrical field.

In a typical experiment a sequence of, equally spaced, charged metal plates is used to generate an electrical field through which electrons are passed (figure 2.5). The electrical fields between the plates are uniform, so the force acting on the electrons between the plates is constant. The energy transferred is then force × distance, the distance being related to the number of pairs of plates through which the electrons pass.

The number of plates connected to the power supply can be varied, so the distance over which the electrons are accelerated can be adjusted. Having passed through the plates, the electrons coast through a fixed distance past two electronic timers that record their time of flight— hence their speed can be calculated. The whole of the interior of the experimental chamber is evacuated so the electrons do not lose any energy by colliding with atoms during their flight.

The prediction from Newtonian physics is quite clear. The work done on the electrons depends on the distance over which they are accelerated, the force being constant. If we assume that the electrons start from rest, then doubling the length over which they are accelerated by the electrical force will double the energy transferred, so doubling the kinetic energy. As kinetic energy depends on v^2, doubling the kinetic energy should also double v^2 for the electrons.

Figure 2.6 shows the variation of v^2 with the energy transferred to the electrons.

Aspects of the theory of relativity

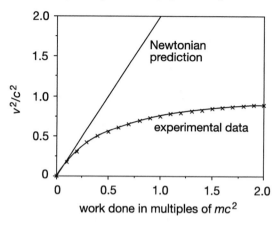

Figure 2.6. Kinetic energy related to work done.

Again the Newtonian result is very wrong. A check can be added to make sure that all the energy is actually being transferred to the electrons. The electrons can be allowed to strike a target at the end of their flight path. From the increase in the temperature of the target, the energy carried by the electrons can be measured. (In order for there to be a measurable temperature rise the experiment must run for some time so that many electrons strike the target.)

Such a modification shows that *all* the energy transferred is passing to the electrons, but the increase in the kinetic energy is not reflected in the expected increase in the velocity of the electrons. Kinetic energy does not depend on velocity in the manner predicted by Newtonian mechanics.

Einstein's theory predicts the correct relationship to be:

$$T = \frac{mc^2}{\sqrt{1 - v^2/c^2}} - mc^2 \tag{2.8}$$
$$= (\gamma - 1)mc^2$$

using the standard abbreviation

$$\gamma = 1/\sqrt{1 - v^2/c^2}.$$

The Newtonian kinetic energy, T, is a good approximation to the true relativistic kinetic energy, T, at velocities much smaller than light. We have been misled about kinetic energy in the same way as we were about momentum. Our *approximately* correct formula has led us into believing that it was the *absolutely* correct one. When particles start to move at velocities close to that of light, we must calculate their kinetic energies by using equation (2.8).

2.3 Energy

There is a curiosity associated with equation (2.8) that may have escaped your notice. One of the first things that one learns about energy in elementary physics is that absolute amounts of energy are unimportant, it is only *changes* in energy that matter[5]. If we consider equation (2.8) in more detail, then we see that it is composed of two terms:

$$T = \gamma mc^2 - mc^2 \qquad (2.8)$$

The first term contains all the variation with velocity (remember that γ is a function of velocity) and the second term is a fixed quantity that is always subtracted. If we were to accelerate a particle so that its kinetic energy increased from T_1 to T_2, say, then the change in kinetic energy (KE) would be:

$$T_1 - T_2 = (\gamma_1 mc^2 - mc^2) - (\gamma_2 mc^2 - mc^2)$$
$$= \gamma_1 mc^2 - \gamma_2 mc^2.$$

The second constant term, mc^2, has no influence on the change in KE. If we were to define some quantity E by the relationship:

$$E = \gamma mc^2 \qquad (2.9)$$

then the change in E would be the same as the change in KE.

The two quantities T and E, differ only by a fixed amount mc^2. T and E are related by:

$$E = T + mc^2 \qquad (2.10)$$

an obvious, but useful, relationship.

We can see from equation (2.10) that E is not zero even if T is. E is referred to as the *relativistic energy* of a particle. The relativistic energy is composed of two parts, the kinetic energy, T, and another term mc^2 that is not zero even if the particle is at rest. I shall refer to this second term as the *intrinsic energy* of the particle[6]. This intrinsic energy is related to the mass of the particle and so cannot be altered without turning it into a different particle.

2.4 Energy and mass

If $E = T + mc^2$, then a stationary particle ($T = 0$) away from any other objects (so it has no potential energy either) still has intrinsic energy mc^2. This strongly suggests that the intrinsic energy of a particle is deeply related to its mass. Indeed one way of interpreting this relationship is to say that all forms of energy have mass (i.e. energy can be weighed!). A metal bar that is hot (and so has thermal energy) would be slightly more massive than the same bar when cold[7].

Perhaps a better way of looking at it would be to say that energy and mass are different aspects of the same thing—relativity has blurred the distinction between them.

However one looks at it, one cannot escape the necessity for an intrinsic energy. Relativity has told us that this energy must exist, but provides no clues to what it is! Fortunately particle physics suggests an answer.

Consider a proton. We know that protons consist of quarks, specifically a uud combination. These quarks are in constant motion inside the proton, hence they have some kinetic energy. In addition, there are forces at work between the quarks—principally the strong and electromagnetic forces—and where there are forces there must be some potential energy (PE). Perhaps the intrinsic energy of the proton is simply the KE and PE of the quarks inside it?

Unfortunately, this is not the complete answer. There are two reasons for this:

(1) Some particles, e.g. the electron, do not have other particles inside them, yet they have intrinsic energy. What is the nature of this energy?

(2) The quarks themselves have mass, hence they have intrinsic energy of their own. Presumably, this quark energy must also contribute

to the proton's intrinsic energy—indeed we cannot get the total intrinsic energy of the proton without this contribution. But then we are forced to ask the nature of the intrinsic energy of the quarks! We have simply pushed the problem down one level.

We can identify the contributions to the intrinsic energy of a composite particle:

$$\text{intrinsic energy of proton} = \text{KE of quarks} + \text{PE of quarks}$$
$$+ \text{intrinsic energy of quarks} \quad (2.11)$$

but we are no nearer understanding what this energy is if the particle has no internal pieces. Unfortunately, the current state of research has no conclusive answer to this question.

Of course, there are theories that have been suggested. The most popular is the so-called 'Higgs mechanism'. This theory is discussed in some detail later (section 11.2.2). If the Higgs theory is correct, then a type of particle known as a Higgs boson should exist. Experimentalists have been searching for this particle for a number of years. As our accelerators have increased in energy without finding the Higgs boson, so our estimations of the particle's own mass have increased. Late last year (2000) the first encouraging hints that the Higgs boson had finally been seen were produced at the LEP accelerator at CERN. We will now have to wait until LEP's replacement the LHC comes on line in 2005 before CERN can produce any more evidence. In the meantime the American accelerator at Fermilab is being tuned to join the search.

2.4.1 Photons

The nature of the photon is an excellent illustration of the importance of the relativistic energy, rather than the kinetic energy. Photons are bursts of electromagnetic radiation. When an individual atom radiates light it does so by emitting photons. Some aspects of a photon's behaviour are very similar to that of particles, such as electrons. On the other hand, some aspects of a photon's behaviour are similar to that of waves. For example, we can say that a photon has a wavelength! We shall see in the next chapter that electrons have some very odd characteristics as well.

Photons do not have electrical charge, nor do they have any mass. In a sense they are pure kinetic energy. In order to understand how this can be, consider equation (2.12), which can be obtained, after some algebraic manipulation, from the relativistic mass and energy equations:

$$E^2 = p^2c^2 + m^2c^4. \tag{2.12}$$

From this equation one can see that a particle with no mass can still have relativistic energy. If $m = 0$, then

$$E^2 = p^2c^2 \quad \text{or} \quad E = pc. \tag{2.13}$$

If a particle has no mass then it has no intrinsic energy. However, as the relativistic energy is not *just* the intrinsic energy, being massless does not prevent it from having *any* energy. Even more curiously, having no mass does not mean that it cannot have momentum. From equation (2.13) $p = E/c$. Our new expanded understanding of momentum shows that an object can have momentum even if it does not have mass—you can't get much further from Newtonian $p = mv$ than that!

The photon is an example of such a particle. It has no intrinsic energy, hence no mass, but it does have kinetic (and hence relativistic) energy and it does have momentum.

However, at first glance this seems to contradict the relativistic momentum equation, for if

$$p = \frac{mv}{\sqrt{1 - v^2/c^2}} \tag{2.4}$$

then $m = 0$ implies that $p = 0$ as well. However there is no contradiction in one specific circumstance. If $v = c$ (i.e. the particle moves at the speed of light) then

$$p = \frac{mv}{\sqrt{1 - v^2/c^2}}$$
$$= \frac{0 \times c}{\sqrt{1 - c^2/c^2}} = \frac{0}{0}$$

an operation that is not defined in mathematics. In other words, equation (2.4) does not apply to a massless particle moving at the speed of light. This loophole is exploited by nature as photons, gluons[8] and possibly neutrinos are all massless particles and so must move at the speed of light.

In 1900 Max Planck suggested that a photon of wavelength λ had energy E given by:

$$E = \frac{hc}{\lambda} \qquad (2.14)$$

where $h = 6.63 \times 10^{-34}$ J s. We now see that this energy is the relativistic energy of the photon, so we can also say that:

$$p = \frac{h}{\lambda} \qquad (2.15)$$

which is the relativistic momentum of a photon and all massless particles.

2.4.2 Mass again

This section can be omitted on first reading.

I would like to make some final comments regarding energy and mass. The equation $E = mc^2$ is widely misunderstood. This is partly because there are many different ways of looking at the theory of relativity. Particle physicists follow the sort of understanding that I have outlined. The traditional alternative is to consider equation (2.4) to mean that the mass of a particle varies with velocity, i.e.

$$p = Mv \qquad \text{where} \quad M = \frac{m}{\sqrt{1 - v^2/c^2}}.$$

In this approach, m is termed the *rest mass* of the particle. This has the advantage of explaining *why* Newtonian momentum is wrong—Newton did not know that mass was not constant.

Particle physicists do not like this. They like to be able to identify a particle by a *unique* mass, not one that is changing with speed. They also point out that the mass of a particle is not easily measured while it is moving, *only the momentum and energy of a moving particle are important quantities*. They prefer to accept the relativistic equations for energy and momentum and only worry about the mass of a particle when it is at rest. The equations are the same whichever way you look at them.

However, confusion arises when you start talking about $E = mc^2$: is the 'm' the rest mass or the mass of the particle when it is moving?

Particle physicists use $E = mc^2$ to calculate the relativistic energy of a particle at rest—in other words the intrinsic energy. If the particle is moving, then the equation should become $E = T + mc^2 = \gamma mc^2$. In other words they reserve 'm' to mean *the* mass of the particle.

2.4.3 Units

The SI unit of energy is the *joule* (defined as the energy transferred when a force of 1 newton acts over a distance of 1 metre). However, in particle physics the joule is too large a unit to be conveniently used (it is like measuring the length of your toenails in kilometres—possible, but far from sensible), so the *electron-volt* (eV) is used instead. The electron-volt is defined as the energy transferred to a particle with the same charge as an electron when it is accelerated through a potential difference (voltage) of 1 V.

The formula for energy transferred by accelerating charges through voltages is:

$$E = QV$$

where E = energy transferred (in joules), Q = charge of particle (in coulombs), V = voltage (in volts), so for the case of an electron and 1 V the energy is:

$$E = 1.6 \times 10^{-19} \text{ C} \times 1 \text{ V}$$
$$= 1.6 \times 10^{-19} \text{ J}$$
$$= 1 \text{ eV}.$$

This gives us a conversion between joules and electron-volts. In particle physics the eV turns out to be slightly too small, so we settle for GeV (G being the abbreviation for giga, i.e. 10^9).

Particle physicists have also decided to make things slightly easier for themselves by converting all their units to multiples of the velocity of light, c. In the equation:

$$E^2 = p^2 c^2 + m^2 c^4 \tag{2.12}$$

every term must be in the same units, that of energy (GeV). Hence to make the units balance momentum, p, is measured in GeV/c and mass is measured in GeV/c^2. This seems quite straightforward until you realize *that we do not bother to divide by the actual numerical value of c.* This can be quite confusing until you get used to it. For example a proton at rest has intrinsic energy = 0.938 GeV and mass 0.938 GeV/c^2, *the same numerical value in each case.* You have to read the units carefully to see what is being talked about!

The situation is sometimes made worse by the value of c being set as one and all the other units altered to come into line. Energy, mass and momentum then all have the same unit, and equation (2.12) becomes:

$$E^2 = p^2 + m^2. \tag{2.16}$$

This is a great convenience for people who are used to working in 'natural units', but can be confusing for the beginner. In this book we will use equation (2.12) rather than the complexities of 'standard units'.

2.5 Reactions and decays

2.5.1 Particle reactions

Many reactions in particle physics are carried out in order to create new particles. This is taking advantage of the link between intrinsic energy and mass. If we can 'divert' some kinetic energy from particles into intrinsic energy, then we can create new particles. Consider this reaction:

$$p + p \rightarrow p + p + \pi^0. \tag{2.17}$$

This is an example of a *reaction equation*. On the left-hand side of the equation are the symbols for the particles that enter the reaction, in this case two protons, p. The '+' sign on the left-hand side indicates that the protons came within range of each other and the strong force triggered a reaction. The result of the reaction is shown on the right-hand side. The protons are still there, but now a new particle has been created as well—a neutral pion, or pi zero, π^0.

The arrow in the equation represents the boundary between the 'before' and 'after' situations.

This equation also illustrates the curious convention used to represent a particle's charge. The '0' in the pion's symbol shows that it has a zero charge. The proton has charge of +1 so it should be written 'p$^+$'.

However, this has never caught on and the proton is usually written as just p. Most other positively charged particles carry the '+' symbol (for example the positive pion or π^+). Some neutral particles carry a '0' on their symbols and some do not (for example the Ξ^0 and the n). Generally speaking, the charge is left off when there is no chance of confusion— there is only one particle whose symbol is p, but there are three different π particles so in the latter case the charges must be included.

All reactions are written as reaction equations following the same basic convention:

$$A + B \rightarrow C + D$$

$$\text{A hits B} \overset{(\text{becoming})}{\rightarrow} \text{C and D.}$$

Reaction (2.17) can only take place if there is sufficient *excess* energy to create the intrinsic energy of the π^0 according to $E = mc^2$. I have used the term 'excess energy' to emphasize that only part of the energy of the protons can be converted into the newly created pion.

One way of carrying out this reaction is to have a collection of stationary protons (liquid hydrogen will do) and aim a beam of accelerated protons into them. This is called a fixed target experiment. The incoming protons have momentum so even though the target protons are stationary, the reacting combination of particles contains a net amount of momentum. Hence the particles leaving the reaction must be moving in order to carry between them the same amount of momentum, or the law of conservation would be violated. If the particles are moving then there must be some kinetic energy present after the reaction. This kinetic energy can only come from the energy of the particles that entered the reaction. Hence, not all of the incoming energy can be turned into the intrinsic energy of new particles, the rest must be kinetic energy spread among the particles:

$$2 \times (\text{protons KE} + \text{IE}) = 2 \times (\text{protons KE} + \text{IE}) + (\pi^0\text{'s KE} + \text{IE})$$

where KE = kinetic energy and IE = internal energy. If the reacting particles are fired at each other with the same speed then the net momentum is zero initially. Hence, all the particles after the reaction could be stationary. This means that all the initial energy can be used to materialize new particles. Such 'collider' experiments are more energy efficient than fixed target experiments. They do have disadvantages as well, which we shall discuss in chapter 10.

In what follows I shall assume that the energy of the incoming particles is always sufficient to create the new particles and to conserve momentum as well.

2.5.2 Particle decays

Many particles are unstable, meaning that they will decay into particles with less mass. We would write such an event as:

$$A \rightarrow B + C$$

A being the decaying particle, and B and C the decay products (or daughters).

In such decays energy and momentum are also conserved. From the point of view of the decaying particle it is at rest when it decays, so there is no momentum. The daughter particles must then be produced with equal and opposite momenta (if there are only two of them). For example the π^0 decays into two photons:

$$\pi^0 \rightarrow \gamma + \gamma.$$

We can use the various relativistic equations to calculate the energy of the photons that are emitted.

From the decaying particle's point of view it is stationary, so only its intrinsic energy will be present initially. After the decay, the total energy of the two photons must be equal to the initial energy, i.e.

$$\text{intrinsic energy of } \pi^0 = \text{total energy of photons}$$
$$M_{\pi^0} c^2 = 2E$$
$$\text{mass of } \pi^0 = 0.134 \text{ GeV}/c^2$$
$$\therefore \qquad 2E = (0.134 \text{ GeV}/c^2) \times c^2$$
$$E = 0.067 \text{ GeV}.$$

The situation is slightly more complicated if the produced particles have mass as well. The best example of this situation is the decay of the K^0. It decays into two of the charged pions:

$$K^0 \rightarrow \pi^+ + \pi^-.$$

To calculate the energy of the pions we follow a similar argument. As the pions have the same mass as each other they are produced with the same energy and momentum. This makes the calculation quite simple.

$$\text{intrinsic energy of K} = \text{total energy of pions}$$
$$M_{K^0}c^2 = 2E_\pi$$
$$\therefore \qquad E_\pi = \frac{M_{K^0}c^2}{2}$$

as the mass of the $K^0 = 0.498 \text{ GeV}/c^2$

$$E_\pi = 0.249 \text{ GeV}.$$

We can go on to calculate the momenta of the produced pions:

$$E^2 = p^2c^2 + m^2c^4$$
$$(0.249 \text{ GeV})^2 = p^2c^2 + (0.140 \text{ GeV}/c^2)^2c^4$$
$$\therefore \qquad p^2c^2 = (0.249 \text{ GeV})^2 - (0.140 \text{ GeV}/c^2)^2c^4$$
$$= 0.042 \text{ (GeV)}^2$$
$$\therefore \qquad p = 0.206 \text{ GeV}/c.$$

2.6   Summary of chapter 2

- Momentum can be measured *directly* by magnetic bending;
- the formula $p = mv$ is only an approximation to the true relationship $p = \gamma mv$;
- relativistic momentum is the quantity that is really conserved in collisions;
- relativistic kinetic energy is given by $T = \gamma mc^2 - mc^2$;
- relativistic energy (a more useful quantity) is $E = \gamma mc^2$;
- mc^2 is the intrinsic energy of a particle which is responsible for giving the particle mass;
- particles that are massless must move at the speed of light;
- such particles (e.g. the photon and the neutrinos) can have energy and momentum even though they do not have mass;
- $E^2 = p^2c^2 + m^2c^4$;
- if $m = 0$, then $p = E/c$ and $E = pc$;
- when particles react excess energy can be diverted into intrinsic energy of new particles, subject to constraints imposed by conservation of energy and momentum;
- when particles decay energy and momentum are also conserved.

Notes

[1] Equation (2.3) is only true if the magnetic field is at $90°$ to the direction in which the particle is moving.

[2] The data are taken from the experiments of Kaufmann (1910), Bucherer (1909) and Guye and Lavanchy (1915).

[3] Consider what happens when the speed of the particle, v, is quite small compared with light. In that case the factor $v^2/c^2 \ll 1$ so that $(1 - v^2/c^2) \approx 1$ and $\sqrt{1 - v^2/c^2} \approx 1$, in which case the relativistic momentum $p \approx mv$!

[4] Equation (2.7) assumes that the electron started its acceleration from rest.

[5] You may think that KE is an exception to this. After all, if an object is not moving then it has no KE, and there cannot be much dispute about that! But consider: any object that is not moving from one person's point of view, may well be moving from another person's viewpoint. Which is correct?

[6] There are various alternative names such as 'rest energy', but they are rooted in the idea of 'rest mass' which is a way of looking at relativity that is not generally used by particle physicists.

[7] The difference in mass is minute as c^2 is a very large number.

[8] Gluons are the strong force equivalent of photons.

Chapter 3

Quantum theory

In this chapter we shall consider some aspects of quantum mechanics. The aim will be to outline the parts of quantum mechanics that are needed for an understanding of particle physics. We shall explore two experiments and their interpretation. The version of quantum mechanics that we shall develop (and there are many) will be that of Richard Feynman as this is the most appropriate for understanding the fundamental forces.

'...*This growing confusion was resolved in 1925 or 1926 with the advent of the correct equations for quantum mechanics. Now we know how the electrons and light behave. But what can I call it? If I say they behave like particles I give the wrong impression; also if I say they behave like waves. They behave in their own inimitable way, which technically could be called a quantum mechanical way. They behave in a way that is like nothing that you have ever seen before. Your experience with things that you have seen before is incomplete. The behaviour of things on a very small scale is simply different. An atom does not behave like a weight hanging on a spring and oscillating. Nor does it behave like a miniature representation of the solar system with little planets going around in orbits. Nor does it appear to be like a cloud or fog of some sort surrounding the nucleus. It behaves like nothing you have ever seen before.*

There is one simplification at least. Electrons behave in this respect in exactly the same way as photons; they are both screwy, but in exactly the same way...

The difficulty really is psychological and exists in the perpetual torment that results from your saying to yourself 'but how can it really be like that?' which is a reflection of an uncontrolled but vain desire to see it in terms of something familiar. I will not describe it in terms of an analogy with something familiar; I will simply describe it...

I am going to tell you what nature behaves like. If you will simply admit that maybe she does behave like this, you will find her a delightful and entrancing thing. Do not keep saying to yourself, if you can possibly avoid it, 'but how can it be like that?' because you will get 'down the drain', into a blind alley from which nobody has yet escaped. Nobody knows how it can be like that.'
From *The Character of Physical Law* by Richard P Feynman.

The 20th century has seen two great revolutions in the way we think about the physical world. The first, relativity, was largely the work of one man, Albert Einstein. The second, quantum theory, was the pooled efforts of a wide range of brilliant physicists.

The period between 1920 and 1935 was one of great turmoil in physics. A series of experiments demonstrated that the standard way of calculating, which had been highly successful since the time of Newton, did not produce the right answers when it was applied to atoms. It was as if the air that physicists breathe had been taken away. They had to learn how to live in a new atmosphere. Imagine playing a game according to the rules that you had been using for years, but suddenly your opponent started using different rules—and was not telling you what they were.

Gradually, thanks to the work of Niels Bohr, Werner Heisenberg, Max Born, Erwin Schrödinger and Paul Dirac (amongst others) the new picture started to emerge. The process was one of continual debate amongst groups of scientists, trying out ideas and arguing about them until a coherent way of dealing with the atomic world was produced.

Contrast this with the development of relativity: one man, working in a patent office in a burst of intellectual brilliance producing a series of papers that set much of the agenda for 20th century physics.

The history of the development of quantum mechanics is a fascinating human story in its own right. However, in this book we must content ourselves with developing enough of the ideas to help us understand the nature of the fundamental forces.

In the chapter on relativity I was able to concentrate on the results of two experiments that demonstrated some of the new features of the theory. In this chapter we shall also look at two experiments (out of historical context). However, much of the time we will be looking at how we interpret what the experiments mean to develop a feel for the bizarre world that quantum mechanics reveals to us.

3.1   The double slot experiment for electrons

A classic experiment that cuts to the heart of quantum mechanics is the double slot experiment for electrons. It is a remarkably simple experiment to describe (although somewhat harder to do in practice) considering how fundamentally it has shaken the roots of physics.

The idea is to direct a beam of electrons at a metal screen into which have been cut two rectangular slots. The slots are placed quite close to each other, closer than the width of the beam, and are very narrow (see figure 3.1). The electron beam can be produced in a very simple way using equipment similar to that found inside the tube of an ordinary television. However, an important addition is the ability to reduce the intensity of the beam so that the number of electrons per second striking the screen can be carefully controlled.

On the other side of the screen, and some distance away, is a device for detecting the arrival of electrons. This can be a sophisticated electronic device such as used in particle physics experiments, or it can be a simple photographic film (of very high sensitivity). The purpose of the experiment is to count the number of electrons arriving at different points after having passed through the slots in the screen.

The experiment is left running for some time and then the photographic film is developed. The exposed film shows a pattern of dark and light patches. In the negative, the dark patches are where electrons have arrived and exposed the film. The darker the patch is the greater the number of electrons that have arrived there during the experiment. A series of such exposures can be used to estimate the probability of

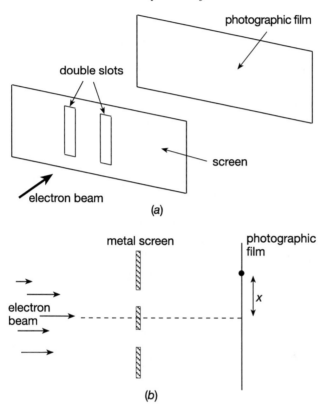

Figure 3.1. (*a*) The double slot experiment for electrons. (*b*) The experiment seen from above.

an electron arriving at any given point (such as that marked *x* on the diagram).

This is a very significant experiment as the results disagree totally with one's expectations. Common sense would have us argue along the following lines[1]:

- an electron will either strike the screen, in which case it will be absorbed, or pass through one of the slots (we can prevent the screen from charging up as electrons hit it by connecting it to the ground);

- any electron that reaches the film on the other side of the screen must have passed through one or the other slot (the screen is so big compared to the beam that none can leak round the edge);
- if the film is 'uniform', i.e. every part of the film is equally sensitive to electrons arriving, and no electrons are lost in transit (the experiment can be done in a vacuum to prevent electrons colliding with atoms in the air) then the film should record the arrival of every electron that gets through a slot;
- so if we could in some way 'label' the electrons as having come through slot 1 or slot 2, then the total number arriving at x will be the sum of those that arrived having come through slot 1 plus those that arrived after having passed through slot 2.

Now we cannot label electrons, but what we can do is to carry out the experiment with one or other of the slots blocked off and count the number of electrons arriving at x. In other words:

$$\begin{array}{ccc} \text{fraction of electrons arriving} \\ \text{at } x \text{ with both slots open} \end{array} = \begin{array}{c} \text{fraction arriving} \\ \text{with slot 1 open} \end{array} + \begin{array}{c} \text{fraction arriving} \\ \text{with slot 2 open} \end{array}$$

The fraction of electrons arriving at a point is the experimental way in which the probability of arriving is measured. So, using the symbol $P_1(x)$ to mean the probability of an electron arriving at x on the screen after having passed through slot 1, etc, we can write:

$$P_{12}(x) = P_1(x) + P_2(x).$$

This is an experimental prediction that can be tested. It may strike you that it is such an obvious statement that there is no need to test it. Well, unfortunately it turns out to be wrong!

Figure 3.2 shows the distribution of electrons arriving at a range of points as measured with one or the other slots blocked off.

Both of these curves are exactly what one would expect after some thought. Most of the electrons that pass straight through arrive at the film at a point opposite the slot. They need not travel through the slot exactly parallel to the centre line (and some might ricochet off the edge of the slot) so the distribution has some width[2].

Now consider the experimental results obtained from carrying out the experiment with both slots open at the same time. This is shown in figure 3.3.

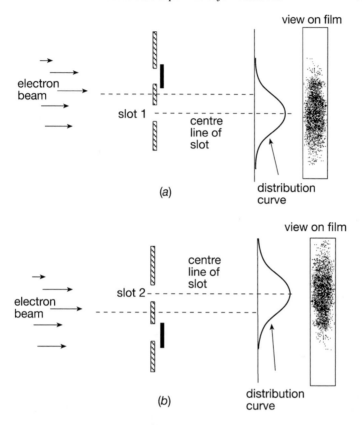

Figure 3.2. (*a*) The electron distribution for slot 1 open and slot 2 closed; the height of the curve indicates the fraction of electrons arriving at that point. (*b*) The distribution of electrons with slot 2 open and slot 1 closed.

No matter how one looks at this result, it is *not* the sum of the other two distributions. Consider in particular a point such as that labelled *y* on the diagram. Comparing the number of electrons arriving here when there are both slots open with the number when only slot 1 is open leads to the odd conclusion *that opening the second slot has reduced the number of electrons arriving at this point*! Indeed points on the film such as *y* show no darkening—electrons *never* arrive there[3].

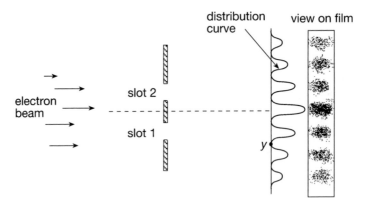

Figure 3.3. The distribution of electrons with both slots open.

As if that were not enough to cause a puzzle, consider the experiment from the point of view of an electron passing through slot 1. If this is the path followed by the electron, then how can it 'know' if slot 2 is open or closed? Yet it clearly can, in some sense. If slot 2 is open then electrons never arrive at a point such as y, yet if slot 2 is closed electrons frequently arrive at y.

As a final twist it is possible to reduce the intensity of the electron beam to such an extent that the time between electrons arriving at the screen is very much greater than the time it takes them to cross from the screen to the film. This ensures that only one electron at a time is passing through the equipment. We would want to do this to ensure that one electron passing through one slot does not have some effect on another passing through the other slot[4].

If we decide to develop the film after each electron arrives, then we should be able to trace the behaviour of a single electron (it is not possible to do this with film, but it can be done with electronic equipment). Unsurprisingly, every time we develop the film we find a single small spot corresponding to the arrival point of a single electron. However, if we stack the films on top of one another, then together all the single electrons draw out the distribution shown in figure 3.3.

You must let the significance of this result sink in. A single electron crossing the equipment will never arrive at one of the zero points in the distribution of figure 3.3. Yet common sense would dictate that a single electron must either go through slot 1 or slot 2 and so should arrive at some point with a relative probability given by either figure 3.2(*a*) or figure 3.2(*b*). Experimentally this has been shown to be incorrect. Every individual electron crossing the equipment 'knows' that both slots are open.

3.2   What does it all mean?

While the experimental facts outlined in the previous section have been established beyond doubt, their interpretation is very much an open question.

When physicists try to construct a theory from their experimental observations they are primarily concerned with two things:

- how to calculate something that will enable them to predict the results of further experiments; and
- how to understand what is going on in the experiment.

Now, it is not always possible to satisfy both of these concerns. Sometimes we have a reasonably clear idea of what is going on, but the mathematical details are so complex that calculating it is very hard. Sometimes the mathematics is quite clear, but the understanding is difficult.

Quantum mechanics falls into the latter category. As the pioneers in this field struggled with the results of experiments just as baffling as the double slot experiment, a very pragmatic attitude started to form. They began to realize that an understanding of *how* electrons could behave in such a manner would come once they developed the rules that would enable them to calculate the way they behaved. So quantum mechanics as a set of mathematical rules was developed. It has been tested in a huge number of experiments since then and has never been found to be wrong. Quantum mechanics as a theory works extremely well. Without it the microchip would not have been developed. Yet we are still not sure how electrons can behave in such a manner.

In some ways the situation has got worse over the years. Some people held out the hope that quantum mechanics was an incomplete theory and that as we did more experiments we would discover some loose end that would allow us to make sense of things. This has not happened. Indeed versions of the double slot experiment have been carried out with photons and, more recently, sodium atoms—with the same results.

If we accept that this is just the way things are and that, like it or not, the subatomic world does behave in such a manner then another question arises. Why do we not see larger objects, such as cricket balls, behaving in this manner? One could set up an experiment similar to the double slot (Feynman has suggested using machine gun bullets and armour plating!), yet one cannot believe that large objects would behave in such a manner. Indeed, they do not. To gain some understanding of why the size of the object should make such a difference, we need to develop some of the machinery of quantum mechanics.

3.3 Feynman's picture

In the late 1940s Richard Feynman decided that he could not relate to the traditional way of doing quantum mechanics and tried to rethink things for himself. He came up with a totally new way of doing quantum mechanical calculations coupled with a new picture of how to interpret the results of the experiments. His technique is highly mathematical and very difficult to apply to 'simple' situations such as electrons in atoms. However, when applied to particles and the fundamental forces the technique comes into its own and provides a beautifully elegant way of working. The Feynman method is now almost universally applied in particle physics, so it is his ideas that we shall be using rather than the earlier view developed by Heisenberg, Schrödinger and Bohr[5].

The starting point is to reject the relationship:

$$P_{12}(x) = P_1(x) + P_2(x).$$

After all, P_1 and P_2 have been obtained from two different experiments and, despite the urgings of common sense, there is no reason in principle why they should still apply in a different, third, experiment. Indeed, evidence suggests that they don't! Accordingly it is necessary to calculate, or measure, the probability for *each* experiment and to be very careful about carrying information from one to another.

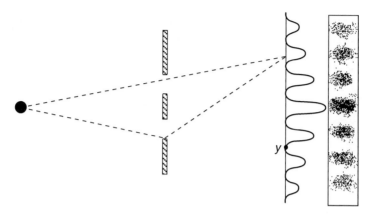

Figure 3.4. Two possible paths leading to the same point on the distribution.

In the first instance quantum mechanics does not work with probabilities, it works with mathematical quantities called *amplitudes*. Amplitudes are related to probabilities:

probability of an event = |amplitude for an event|^2.

but they are not the same thing. (The straight brackets, |, indicate that this is not a simple squaring as with numbers, e.g. $2^2 = 4$. This is known as the *absolute square*. The way in which the absolute square is calculated is described later.) Amplitudes cannot be measured directly. Only probabilities can be measured. However, it is the amplitudes that dictate the motion of particles.

The basic rules of quantum mechanics are quite clear about this. To calculate the probability of an event happening (an electron arriving at a point on a film) one must calculate the amplitude for it to happen, and then absolute square the result to get the probability. Why not just calculate the probability directly? Because amplitudes *add* differently to probabilities.

According to Feynman, to calculate the amplitude for an electron arriving at a point on a film one has to consider all the possible paths that the electron might have taken to get there. In figure 3.4 a couple of possible paths are drawn.

Feynman invented a method for calculating the amplitude for each possible path.   To get the total amplitude one just adds up the contributions for each path. Writing the amplitude to get from a to b as $A(a, b)$:

$$A(a, b) = A(a, b)_{\text{path 1}} + A(a, b)_{\text{path 2}} + A(a, b)_{\text{path 3}} + \cdots$$

and

$$\text{probability}(a, b) = |A(a, b)_{\text{path 1}} + A(a, b)_{\text{path 2}}$$
$$+ A(a, b)_{\text{path 3}} + \cdots |^2.$$

Now, as it is the total amplitude that is absolute squared the result is not the same as taking the amplitudes separately. Consider the simple case of just two paths:

- amplitude for path 1 $= A_1$
- amplitude for path 2 $= A_2$
- so, probability if only path 1 is available $= |A_1|^2$
- probability if only path 2 is available $= |A_2|^2$
- but, if both paths are available then:
  probability $= |A_1 + A_2|^2 = |A_1|^2 + |A_2|^2 + 2 \times A_1 \times A_2$
  and it is the extra piece in this case that makes the difference.

It does not matter how odd or unlikely the path seems to be, it will have an amplitude and its contribution must be counted. Sometimes this contribution will be negative. That will tend to reduce the overall amplitude. In some cases the amplitude is made up of contributions from different paths that totally cancel each other out (equal positive and negative parts). In this case the amplitude to get from that a to that b is zero—it never happens. Such instances are the zero points in the distribution of figure 3.3. Electrons never arrive at these points as the amplitudes for them to do so along different paths totally cancel each other out. Zero amplitude means zero probability.

Clearly, if one of the slots is blocked off then there are a whole range of paths that need not be considered in the total sum. This explains why the probability distributions look different when only one slot is open.

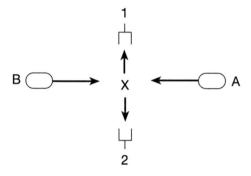

Figure 3.5. Scattering nuclei off each other.

3.4   A second experiment

The second experiment that we need to consider involves scattering particles off one another (see figure 3.5). This will reveal some more interesting features of calculating with amplitudes.

The experimental equipment consists of two sources (placed at A and B) of two different nuclei, α and β. The sources are aimed at each other, and two detectors, 1 and 2, are placed at right angles to the path of the nuclei. The simple idea is to collide the nuclei at point X and count how many arrive at detector 1 and detector 2.

To calculate the probability that a nucleus from A arrives at 1 we need to calculate the amplitude for it to cross from A to X, and then the amplitude for it to go from X to 1. Then these amplitudes must be combined. In this instance the two events, α goes from A to X *then* from X to 1 (we will abbreviate this to $\alpha : A \rightarrow X, X \rightarrow 1$), must happen *in succession*, so the amplitudes *multiply*:

$$\text{Amp}(\alpha : A \rightarrow 1) = \text{Amp}(\alpha : A \rightarrow X) \times \text{Amp}(\alpha : X \rightarrow 1)$$

similarly

$$\text{Amp}(\beta : B \rightarrow 2) = \text{Amp}(\beta : B \rightarrow X) \times \text{Amp}(\beta : X \rightarrow 2).$$

This follows from the rules of ordinary probability. If four horses run in a race the chances of randomly selecting the winner are 1/4. If a second race contains seven horses, then the chances of betting on the winner in that race are 1/7. In total the chance of betting on the winner in both races is $1/4 \times 1/7$ or $1/28$.[6]

Now, if we were calculating the probability of a nucleus ending up at detector 1 and another nucleus at detector 2 (we don't mind which one goes where), then according to the normal rules of probability we would say that:

prob(a nucleus at 1 + a nucleus at 2) = prob(nucleus at 1)

+ prob(nucleus at 2).

However, we might suspect that quantum mechanics uses a different rule for this. It turns out, however that this *does* give us the correct result. It is quite complicated in terms of our amplitudes, as we do not mind which nucleus goes where:

prob(a nucleus at 1 + a nucleus at 2)
$$= |Amp(\alpha : A \rightarrow 1) \times Amp(\beta : B \rightarrow 2)|^2$$
$$+ |Amp(\alpha : A \rightarrow 2) \times Amp(\beta : B \rightarrow 1)|^2$$
$$= 2p$$

(if the probability to go to a detector is the same for each nucleus $= p$).

In this case we do not add all the amplitudes first before we square as the two events we are considering are *distinguishable*. Nucleus α and nucleus β are different, so we can tell in principle (even if we choose not to bother) which one arrives at which detector.

Now we consider a different version of the experiment. This time we arrange that the two nuclei are exactly the same type (say both helium nuclei). According to normal probability rules it should not make any difference to our results. Yet it does. If we now measure the probability of a nucleus arriving at 1 and another arriving at 2, we get a different answer!

This time, we cannot tell—*even in principle*—if it is the nucleus from A that arrives at 1 or the nucleus from B, so we have to combine the amplitudes *before* we square:

$$\text{prob(a nucleus at 1 + a nucleus at 2)}$$
$$= |\text{Amp}(\alpha : A \to 1) \times \text{Amp}(\beta : B \to 2))$$
$$+ (\text{Amp}(\alpha : A \to 2) \times \text{Amp}(\beta : B \to 1)|^2$$
$$= 4p$$

a result that is confirmed by doing the experiment.

Classically these two experiments would yield the same result. Classically we would expect that two nuclei will scatter off each other and be detected in the same manner, their being identical nuclei should not make a difference[7].

Consider this analogy. Imagine that you have a bag that contains four balls, two of which are white and two of which are black. The balls are numbered—1 and 2 are black, 3 and 4 are white—so you can tell which balls are being picked out. If you pick out a ball, replace it and then pick out another, the following possibilities are all equally likely:

1B + 1B	1B + 2B	1B + 3W	1B + 4W
2B + 2B	2B + 1B	2B + 3W	2B + 4W
3W + 3W	3W + 1B	3W + 2B	3W + 4W
4W + 4W	4W + 1B	4W + 2B	4W + 3W

which gives the probability of picking out a white ball and a black ball (it doesn't matter which ones—all the underlined are possibilities) as being 8/16 or 50:50.

If we repeat the experiment having wiped off the numbers what are the chances of picking out a white and black ball? Classically they are the same (and indeed for balls they *are* the same as the quantum effects are small on this scale). One cannot imagine that wiping off the numbers would make a difference. However, according to quantum mechanics it does. Now we cannot tell which ball is which and the possible results are:

$$W + W \qquad W + B \qquad B + B$$

giving a probability of picking out a white and black as 1/3. In other words, if these were quantum balls then the sequences underlined in the first list would all be the same event. It is not simply that we could not *tell* which event we had actually seen, *they are the same event.* Rubbing the numbers off classical balls may make it harder to distinguish them, but it is still possible (using microscopic analysis for example). With the quantum balls the rules seem to be telling us that their identities merge.

Summarizing, the rules for combining amplitudes are:

> if the events cannot be distinguished, amplitudes add before squaring;
> if the events can be distinguished, square the amplitudes then add;
> if the events happen in a sequence, multiply the amplitudes.

This relates to the calculation of path amplitudes in the double slot experiment. With the equipment used in the experiment, there is no way to tell which path the electron passes along, so they must all be treated as indistinguishable and their amplitudes added.

3.5   How to calculate with amplitudes

In this section the mathematical rules for combining amplitudes are discussed. Readers with a mathematical bent will gain some insight from this section. Readers without a taste for such matters will not lose out when it comes to the later sections, provided they read the summary at the end of this section.

Amplitudes are not numbers in the ordinary sense, which is why they cannot be measured. Ordinary numbers represent quantities that can be counted (5 apples on a tree, 70 equivalent kilogram masses in a person, 1.6×10^{-19} Coulombs charge on a proton). Ordinary numbers have size. Amplitudes have both *size* and *phase*.

Consider a clock face (figure 3.6). The short hand on a clock has a certain length (size). At any moment it is pointing to a particular part of the clock face. By convention, we record the position of the hand by the angle that it has turned to from the straight up (12 o'clock) position. To tell the time we must specify the size (length) and phase (angle) of both hands on the face.

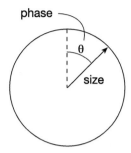

Figure 3.6. The size and phase of a clock hand.

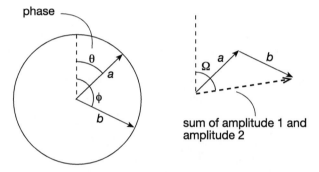

sum of amplitude 1 and
amplitude 2

Figure 3.7. The rule for adding amplitudes is to place them nose to tail; the total amplitude is the line connecting the first tail to the second nose.

The rule for adding quantities that have both size and phase was established by mathematicians long before quantum mechanics was developed (figure 3.7):

amplitude 1: size a, phase θ
amplitude 2: size b, phase ϕ
amplitude 3 = amplitude 1 + amplitude 2
amplitude 3 has size r where $r^2 = a^2 + b^2 + 2ab\cos(\phi - \theta)$
and phase Ω is given by $r\sin(\Omega) = a\sin(\theta) + b\sin(\phi)$.

These are quite complicated rules and we will not be using them in this book. However, note that according to the rule for adding sizes, two

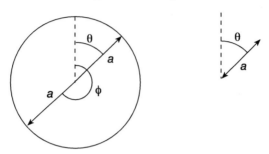

Figure 3.8. These two amplitudes have the same size, but the phases are 180°
different; when they are added together, the total is zero.

amplitudes of the same size can add to make an amplitude of zero size if
their phases are different by 180° (figure 3.8).

If amplitude 1 has size a, and phase θ, and amplitude 2 has size a, and
phase $\phi = 180° + \theta$, then amplitude 3 has size r where

$$
\begin{aligned}
r^2 &= a^2 + b^2 + 2ab\cos(\phi - \theta) \\
&= r^2 + r^2 + 2rr\cos(180°) \\
&= 2r^2 + 2r^2(-1) \\
&= 0 \qquad (\text{as } \cos(180°) = -1).
\end{aligned}
$$

There are also rules for multiplying amplitudes together:

$$
\begin{aligned}
\text{amplitude 1} \times \text{amplitude 2} &= \text{amplitude 3} \\
\text{size of amplitude 3} &= (a \times b) \\
\text{phase of amplitude 3} &= \theta + \phi.
\end{aligned}
$$

The comparative simplicity of these rules suggests that multiplying is a
more 'natural' thing to do with amplitudes than adding them.

If the two amplitudes have the same size, but phases that are opposite
(e.g. 30° and −30°) then when they are multiplied together the size is
squared ($a \times a = a^2$) but the phase is zero. Such pairs of amplitudes
are called *conjugates*, and multiplying an amplitude by its conjugate is
absolute squaring the amplitude. Whenever an amplitude is squared

in this fashion, the phase is always zero. Our earlier rule relating amplitudes to probabilities should more properly read:

$$\text{probability} = \text{amplitude} \times \text{conjugate amplitude} = |\text{amplitude}|^2$$

which is why we never have to deal with the phase of a probability—it is always zero.

Summary

- Amplitudes have both size and phase;
- phase is measured in degrees or radians;
- adding amplitudes is a complex process which results in two amplitudes of the same size adding up to zero if they are 180° different in phase;
- when amplitudes are multiplied together their sizes multiply and their phases add;
- conjugate amplitudes have the same size but opposite phase;
- multiplying an amplitude by its conjugate produces a quantity with a phase of zero—an ordinary number.

3.6 Following amplitudes along paths

Thirty-one years ago Dick Feynman told me about his 'sum over histories' version of quantum mechanics. 'The electron does anything it likes', he said. 'It just goes in any direction at any speed,... however it likes, and then you add up the amplitudes and it gives you the wavefunction'. I said to him, 'You're crazy'. But he wasn't.

Freeman Dyson 1980

Feynman's technique for calculating probabilities in quantum mechanics involves associating an amplitude with every path that a particle might take. It is best to think of a path as a series of numbers representing the position of a particle at a given moment in time. In a one-dimensional case a path would then be a series of number pairs (x, t) between the start (x_a, t_a) and the end (x_b, t_b).

Once the individual path amplitudes have been calculated they are added together to get the final amplitude, which gives you the probability when you take the absolute square. The different paths are assumed to be indistinguishable, which is why the amplitudes are summed before taking the absolute square.

There is one subtlety about this approach that needs a little thought. To calculate an amplitude for a path, the start, (x_a, t_a), and the end, (x_b, t_b), are both needed. If the final answer is to come out right, then the start and end had better be the same for each path *both in terms of the position and the time*. Consequently the particle must travel along each path in the same period of time, $(t_b - t_a)$, so *paths of different lengths must be covered at different speeds*. Anything is allowed. Summing over all the paths really does mean over every possible path—*including all varieties of speed*.

This is such a radical idea that it needs further thought. Common sense would suggest that the electron travels from a to b following a definite (single) path moving at a given speed at any moment. As we have already seen, common sense is not a good guide to what happens in quantum theory and Feynman's way of calculating seems to split from the common sense view in an extreme way. All paths and all speeds must be considered—we cannot hold to a clear picture of any single thing that the electron is 'doing' at any one time. One might be tempted to consider the amplitudes that we calculate as 'representing', in some manner, the electron—yet in Feynman's version of quantum theory this thought is hard to maintain. The amplitude is a mathematical object that relates the probability of a process happening; if it represents the electron in any manner, it is only as a measure of the potentials for change inherent in the situation[8].

We must now consider how Feynman's approach is used to calculate amplitudes. Imagine splitting a path into lots of different segments (as in figure 3.9). In general paths involve three dimensions, but in figure 3.9 only one space direction is shown. The path has been split into equal short intervals of time ε.

Feynman discovered that the amplitudes for the different paths have the same size, but not necessarily the same phase. When a particle advances from position x_1 at time t_1 to position x_2 at time t_2 the phase changes by

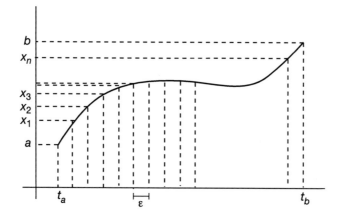

Figure 3.9. A path (here one-dimensional) can be split up into a series of points by specifying times and the position at those times. In this case the time has been split up into n equal intervals ε along the horizontal time axis.

an amount:

$$\text{change in phase between } (x_1, t_1) \text{ and } (x_2, t_2) = \frac{\varepsilon}{\hbar} L$$

where $\varepsilon = (t_2 - t_1)$ is the small time difference, $\hbar$ is Planck's constant divided by 2π and L is a mathematical expression known as the *Lagrangian* of the system (we will see shortly how the Lagrangian is constructed in a simple case). Once the Lagrangian is known, the phase for a given path can be calculated by advancing from one end of the path to the other in small steps using the formula above to obtain the changes in phase for each step and adding them up. Those of a mathematical bent will realize that we are really integrating the Lagrangian along the path and that the phase of the amplitude for that path is given by

$$\text{phase of path} = \frac{1}{\hbar} \int_{t_1}^{t_2} L \, dt.$$

Once this is known, the contributions for different paths can then be added together to give:

$$\begin{array}{c}\text{total phase of amplitude} \\ \text{to get from } x_1 \text{ to } x_2 \end{array} = \sum_{\text{different paths}} \left(\frac{1}{\hbar} \int_{t_1}^{t_2} L \, dt \right).$$

What remains, then, is to specify a Lagrangian and discover how it differs for different paths. For an electron that is moving quite slowly, so we do not have to worry about relativity, an appropriate Lagrangian is:

➡ $$L_{\text{whole path}} = \tfrac{1}{2}mv^2 - V(x, t). \qquad (3.1)$$

This is the difference between the kinetic energy of the particle (the first term) and its potential energy (second term)[9]. The Lagrangian reduces to the following form if we are only worried about a path segment:

➡ $$L_{\text{section of path}} = \frac{1}{2}m\frac{(x_2 - x_1)^2}{\varepsilon^2} - V\left(\frac{x_1 + x_2}{2}, \frac{t_1 + t_2}{2}\right).$$

Note that the speed is calculated for each segment of the path. If you recall the path is specified in terms of a set of positions and times, so that $(x_1 - x_2)/\varepsilon$ is the speed of the particle between two points on the path. The potential energy is calculated at a point mid-way along the path segment.

The Lagrangian is a well known object from classical physics, but Feynman, following up a hint by Dirac, was the first person to see how it could be employed in quantum mechanics.

3.6.1   The Lagrangian in classical physics

At school students learn how objects move when forces are acting on them. For example a cricket ball thrown in from the outfield will follow a curved path. If a student knew the speed of the ball as it left the fielder's hand and the angle at which it was thrown, then he or she could figure out how far the ball would travel and the shape of the curved path. All that would be needed is some further information about how strongly gravity was pulling on the ball.

The basic method for doing this was devised by Newton and works very well for cricket balls that are moving slowly compared with the speed of light. The Newtonian method for figuring out the shape of the path is to use the laws of motion to calculate the speed, position and acceleration of the ball from moment to moment and then to piece the path together bit by bit.

Consider the various curved paths shown in figure 3.10.

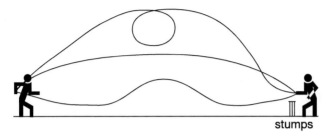

stumps

Figure 3.10. Various paths that might be taken by a thrown cricket ball. Each path has an action that depends on the shape of the path. The actual path followed by the ball is the one with the smallest action.

One of the paths is the 'correct' one—the path that would be followed by the thrown cricket ball. The others are incorrect and if the ball was to travel along one of these paths, it would violate the laws of motion as set down by Newton. Each path can be associated with a quantity called the *action*, *S*, which is calculated from the Lagrangian:

$$S = \int L \, dt. \tag{3.2}$$

To carry out this calculation an equation for the path must be bolted into the formula for the Lagrangian, which for a cricket ball will be the same as equation (3.1) with the potential energy *V* being due to gravity. One could guess the shape of the path and use the equation of that shape to calculate the action. One could even carry on doing this for different shapes until you got bored. It turns out that *the correct path as followed by the cricket ball is the one with the smallest action.*

In practice this is not quite how it is done. There is no need to guess various shapes of path and try each one in turn. The hypothesis that the correct path is the one with the smallest action can be used to derive from (3.2) a different equation that can be solved to give the right path.

This method of figuring out the path of the cricket ball is dramatically different from the traditional Newtonian process. Newton would have you assemble the path by considering the moment-to-moment effect of having a force acting on the ball (in this case the force is gravity). The Lagrangian method is to find the correct path by looking at the shape of the path as a whole, not by breaking it down into pieces.

Which method is easier depends very much on the complexity of the problem; for a cricket ball the Newtonian method works very well. However, there is an important difference in the philosophy behind the two approaches. In many ways the Newtonian method is more intuitive. One could imagine filming the ball in flight and by slowing down a replay seeing how its speed and position changed from instant to instant exactly as predicted by Newton's laws. The Lagrangian approach suggests that somehow the ball has picked out the correct path by having some sort of 'feel' for the path as a whole and that the correct path is the one with the smallest action. This seems to imply that at any moment the ball is 'aware' of the way in which the path is going to evolve as it moves.

3.6.2   Relating classical to quantum

This is starting to sound very similar to the sort of statements that I was making about electrons passing through the double slot experiment. Indeed one of the aspects of his formulation of quantum theory that Feynman was most proud of was the light it shed on the classical Lagrangian way of doing mechanics. According to Feynman the cricket ball actually 'sniffs out' all possible paths at the same time. Each path is given an amplitude according to the action along that path. However, and here is the crunch point, it is only along those paths that are similar to the 'actual' classical path that the phases of the amplitudes will be similar. Paths that stray a long way from the classical path will have widely differing phases (positive and negative) which tend to cancel each other out.

Consider figure 3.11 in which A and B represent two points on the classical path of the cricket ball. They have been connected by several different paths that would be impossible according to Newton's laws. The classically correct path is shown by the dotted line and is labelled 1. Paths that are close to this, such as 2, will have amplitudes with phases that are almost the same as that along the dotted path. Each of these will add to the overall amplitude for getting from A to B. Paths further from the classical, such as 3, will tend to cancel each other out.

So why then do we not see some strangely fuzzy ball zipping through all the mad paths from A to B?

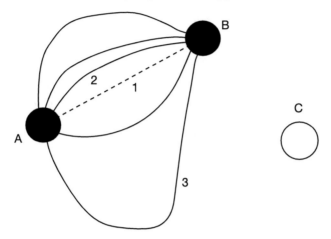

Figure 3.11. The classical path of a ball can be broken into points, such as A and B, connected by many quantum paths.

Partly this is because of the size of the ball, which is very much bigger than that of the electrons, protons etc inside it. They can be zipping along their own simultaneous quantum paths, which are differing from one another by considerable amounts and yet still be inside the ball!

Another thing to remember is that we see the ball simply because photons of light are reflected from it to our eyes. At the quantum level these photons are interacting with the charged particles within the ball—a process that is, once again, governed by various quantum paths. Electron paths close to the classical one have amplitudes that add up and they are also the ones with the biggest amplitudes to reflect the photons to our eyes.

A third point can be made by considering figure 3.11 again. What would be the chance of the cricket ball going from A to C rather than A to B? As the value of $\hbar$ is small (1.05×10^{-34} J s) and the value of the cricket ball's Lagrangian is pretty big, a small change in path makes a very big change to the phase. This means that only paths that are *very* close to the classical one will have similar phases. All the paths connecting A and C together are way off the classical line and so their amplitudes will have a variety of different phases which will add together to give something very close to zero.

3.6.3 Summarizing

The Feynman prescription for quantum mechanics can be summarized as follows:

- every path contributes an amplitude;
- the total amplitude is found by adding up the amplitudes for each path;
- each amplitude has the same size, but the phase depends on the Lagrangian and the shape of the path;
- some paths arrive with similar phases, so they add to increase the total amplitude;
- some paths have amplitudes with phases that are 180° or more apart, so they add to reduce the total amplitude;
- the probability is found by absolute squaring the total amplitude.

The rules are quite clear that every path must be considered in this calculation. If this is not done, then the calculated probability does not agree with the value that can be estimated from experimental results. This means that even the paths that are impossible as judged by classical physics must be taken into account as well.

3.6.4 Amplitudes, energy and momentum

This section involves a fair bit of messing with algebra, so those with a weak mathematical stomach can simply move on to the next section without missing much.

For a free particle (i.e. one that is not acted upon by forces, so there is no potential energy in the Lagrangian) the Lagrangian governing the phase change after a small step is:

➡ $$L = \frac{1}{2}m\frac{(x_2 - x_1)^2}{\varepsilon^2}.$$

If this is used to calculate the amplitude for a particle to move from (x_a, t_a) to (x_b, t_b) by any possible path, then the result has

➡ $$\text{phase} = \frac{\varepsilon L}{\hbar} = \frac{m(x_b - x_a)^2}{2\hbar(t_b - t_a)}.$$

Over short distance (and I will say a little more about what 'short' means in this context in a moment) this is a mess. All the quantum paths contribute significantly to the overall phase. However, if x_b is quite a long way away from x_a, then the phase settles down into a regularly recurring pattern. This pattern turns out to be quite interesting.

For convenience let us make x_a, t_a the origin of space and the start of our timing, i.e. $x_a = 0$ and $t_a = 0$. Let us also suggest that the particle has travelled a long distance X in time T.

Consequently:

➡
$$\text{phase at } X = \frac{mX^2}{2\hbar T}.$$

Now we ask what happens if we move along an *additional* small distance x. This implies that the particle will have been travelling a small extra time t. The phase of the amplitude to get from the origin to $X + x$ is then:

$$\text{phase at } X + x = \frac{m(X + x)^2}{2\hbar(T + t)}.$$

This looks a bit nasty, so we had better tidy it up a little. We can expand the bracket on the top so that we get:

$$\text{phase} = \frac{m(X^2 + x^2 + 2Xx)}{2\hbar(T + t)}.$$

I propose that we ignore the term x^2 on the basis that x is very much smaller than X, so that x^2 will be much smaller than X^2 and also Xx.

$$\text{phase} = \frac{m(X^2 + 2Xx)}{2\hbar(T + t)}.$$

The next step involves worrying about the $(T + t)$ on the bottom. One trick is to write $(T + t)$ as $T(1 + t/T)$ which does not look like much help, but there is an approximation that we can now use. Basically if y is a small number, then:

$$\frac{1}{(1 + y)} = 1 - y.$$

(Try it with a calculator if you don't believe me). Using this we get:

$$\text{phase} = \frac{m(X^2 + 2Xx)}{2\hbar(T + t)} = \frac{m(X^2 + 2Xx)}{2\hbar T(1 + \frac{t}{T})} = \frac{m(X^2 + 2Xx)}{2\hbar T}(1 - t/T)$$

$$= \frac{m(X^2 + 2Xx)}{2\hbar T} - \frac{mt(X^2 + 2Xx)}{2\hbar T^2}.$$

But hang on, you say, this looks *worse* not better. However, we now stop worrying about the phase at $X + x$ and start asking by how much the phase has *changed* in the journey from X to $X + x$. This can be calculated by subtracting the phase at X:

$$\text{phase change} = \left[\frac{m(X^2 + 2Xx)}{2\hbar T} - \frac{mt(X^2 + 2Xx)}{2\hbar T^2} \right] - \frac{mX^2}{2\hbar T}.$$

The term on the far right will kill off part of the first term in the square bracket to give:

$$\text{phase change} = \frac{mXx}{\hbar T} - \frac{mtX^2}{2\hbar T^2} - \frac{mtXx}{\hbar T^2}.$$

Just for a moment I am going to split this up in a different manner to give:

$$\text{phase change} = \frac{mX}{\hbar T} \left[x - \frac{Xt}{2T} - \frac{xt}{T} \right].$$

This shows us that we can drop the last term compared to the other two. The factor t/T is going to make the last couple of terms rather small, but the last one is made even smaller by the fact that x is small compared with X. Having dropped the last term we can repackage what is left:

➡ $$\text{phase change} = \frac{mXx}{\hbar T} - \frac{mtX^2}{2\hbar T^2} = \left(\frac{mv}{\hbar} \right) x - \left(\frac{\frac{1}{2}mv^2}{\hbar} \right) t \quad (3.3)$$

where I have used the fact that, classically, a particle travelling a distance X in time T would be moving at speed $v = X/T$.

We have nearly reached the result I want. For the final step recall that the phase of an amplitude has been compared to a clock hand moving round. Clearly on the face of a clock the phase repeats itself every revolution— once the phase has increased by 2π it basically goes back to the start again.

Our formula shows us how the phase is changing once the particle has travelled a long distance X. If we could keep the time fixed and move along in x, then the phase would repeat itself when $\left(\frac{mv}{\hbar} \right) \times x = 2\pi$ or when x increased by $\frac{2\pi\hbar}{mv} = \frac{h}{mv}$. If this were a classical wave, such as a ripple on the surface of a pond, we would say that the distance between

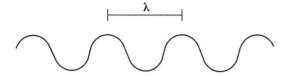

Figure 3.12. The wavelength, λ, of a wave.

two points of equal phase was the *wavelength* of the wave. Applying the same idea we could say that the wavelength of the amplitude was $\frac{h}{mv}$. This is the same relationship as that suggested by Max Planck for the wavelength of a photon (equation (2.15)).

Waves have a period as well as a wavelength. The period is the time you have to wait at a given spot for the phase to repeat itself. For our amplitude the period would be given by $(\frac{1}{2}mv^2/\hbar)t = 2\pi$ or $t = \frac{h}{KE}$ with KE being the kinetic energy of the particle.

After all this messing about with algebra it would not be surprising if the reader had lost the thread of what was going on. Keep hold of this—if a particle has not travelled very far (i.e. a distance that is small compared to a wavelength) then the phase will be a mess and will not follow any fixed pattern as lots of different paths are contributing equally. However, if we let the particle travel a long way (i.e. a distance that is large compared to a wavelength) then the phase will have settled down into a pattern that repeats itself every wavelength (a distance that depends on the classical momentum of the particle) and every period (a time that depends on the classical kinetic energy). These repeats in the amplitude are what people sometimes refer to as the wavelike nature of a quantum particle.

A reasonable question would be to ask what typical wavelengths are. For an electron with $p \sim 60$ keV/c:

$$\lambda = \frac{h}{p} = \frac{6.63 \times 10^{-34}}{3.2 \times 10^{-23}} = 2.1 \times 10^{-11} \text{ m}$$

which is comparable to the size of an atom. Consequently once the electron has moved a distance further than several atomic sizes, we can expect the amplitude to settle into a repeating pattern. However, something the size of a cricket ball travelling at 10 m s^{-1} would have a quantum wavelength of

$$\lambda = \frac{h}{p} = \frac{6.63 \times 10^{-34}}{0.1 \text{ kg} \times 10 \text{ m s}^{-1}} = 6.63 \times 10^{-34} \text{ m}$$

which is far too small for any wave effects to be observed.

3.6.5   Energy levels

It is quite commonly known that electrons inside atoms can only exist with certain values of energy called the *energy levels*. If you try to insert an electron into an atom with a different energy, then it will not fit and either come out of the atom, or radiate away the difference in energy and settle into one of the levels. This is often compared to the rungs on a ladder. One can stand on the rungs, but if you try to stand between the rungs all you do is fall down onto the next level.

Energy levels are a common quantum mechanical consequence of having a closed system—i.e. a system where the paths are localized in space.

Electrons within an atom must stay within the boundary of the atom as they do not have enough energy to escape. If we consider the motion of electrons within atoms as being closed paths (i.e. ones that return to their starting point) then we can calculate the amplitude to go round the atom and back to the start. If we do this, then the phase at the start of the path should be the same as the phase at the end[10]. This must be so as the start and end are the same place and even quantum mechanics does not allow two different phases at the same point! Within a given atom, the Lagrangian is such that only certain energy values allow this to happen. Hence there are energy levels within atoms.

The same is true of other closed systems. Quarks within protons and neutrons have energy levels, however in this case the Lagrangian is much more complicated as the strong force between the quarks has to be taken into account when calculating the potential energy.

3.6.6   Photons and waves

In the previous chapter I mentioned that photons, which in many ways seem like particles, can be assigned a wavelength which is a property traditionally only associated with waves. We can now see, from

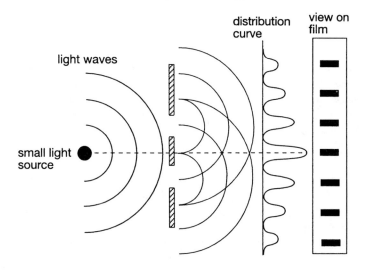

Figure 3.13. A double slot experiment for light.

our previous study of electrons, how this might come about with an appropriate Lagrangian for a photon. We can carry out the double slot experiment using photons rather than electrons and get exactly the same sort of result (see the quotation from Feynman at the start of this chapter).

The double slot experiment was first carried out for light by Young in 1801 (see figure 3.13). It was interpreted as evidence for the wave nature of light. This wave nature is arising out of the underlying effects of the path summations.

The experiment can be explained in classical wave terms in the following manner. When light waves from a small source strike the screen they produce secondary waves centred on the slots. These waves will then arrive at the photographic film having travelled from one slot or the other. Depending on the distance travelled, the waves will arrive either in phase with each other (i.e. peaks or dips on the waves arrive at the same moment) or out of phase (i.e. a dip on one wave arrives at the same moment as a peak on the other). In phase produces re-enforcement and hence a patch of light on the film, out of phase produces cancellation and so a dark region.

However, if this experiment is done in a more sophisticated manner, using a laser, then a puzzling aspect emerges.

If the intensity of the laser is turned right down and the film replaced with electronic detectors, then the pattern of light and dark bands breaks up into a collection of speckles, just like the pattern for individual electrons. This is interpreted as evidence for the laser producing individual photons that cross the experiment one at a time. Yet, just like the electrons, they 'know' about both slots being open.

This is the crux of one of the most bizarre aspects of quantum mechanics. Wave and particle natures seem interchangeable. As we develop the machinery of quantum mechanics more thoroughly over the next few pages, we will start to see how the modern physicist deals with this problem. Further insight will come later, in chapter 11, when we start to study the quantum field.

3.7   Amplitudes, states and uncertainties

3.7.1   The state vector

In our discussion of the double slot experiment and the Feynman method for calculating amplitudes, we were treating the experiment as if we knew where the electron arrived at the screen and then asking what the probability of that arrival point was. It is, of course, equally valid to ask before the experiment is run what the probability of arriving at any point on the screen is. The calculation would proceed as before, but would have to be repeated for every possible arrival point. This would produce a set of amplitudes—one for every point on the screen:

$$[A_1(t), A_2(t), A_3(t), A_4(t), \ldots, A_n(t)]$$

in which A_1 stands for the amplitude that the electron will arrive at point 1 on the screen etc. The amplitudes have also been shown to be dependent on time, $A(t)$, which in general they will be. Mathematicians call an array of numbers like this a *vector*, so in quantum mechanics such a set of amplitudes, $[A_1(t), A_2(t), A_3(t), A_4(t), \ldots, A_n(t)]$, is sometimes rather grandly known as a *state vector*.

In principle, the number of amplitudes even on a small screen is infinite as there are a vast number of microscopic points on the screen. Given this it makes more sense to write the amplitude as $A(x, t)$ with x standing for position on the screen. This makes it clear that the amplitude is a mathematical expression depending on position and time.

Generally the question we are asking is 'what is the probability of an electron being found at point x', in which case we are not too worried about the starting position. The electron may very well have started from somewhere in a region—in other words we may not be sure of the exact starting point. In such circumstances the Feynman method works as well, we just have to deal with every starting position.

An expression such as $A(x, t)$ is also a state vector, but because it is not always a finite set of numbers it is also referred to as a *wave function*[11]. The wave function is the amplitude for finding an electron at any point x and at any time t. We have seen how a wave function can be built up from a set of path sums, but as it is a rather confusing process it is probably best to summarize this again:

amplitude for an electron arriving at point x at time t

$$= \text{constant} \times \exp \left(i \left[\sum_{\text{all starting points}} \left(\sum_{\text{all paths}} \left(\frac{1}{\hbar} \int_{t_1}^{t_2} L \, dt \right) \right) \right] \right).$$

Note that the constant at the start of this formula is to account for the summing of the amplitudes which are of the same size but different phase. Those readers with a taste for mathematics may be interested in the contents of the box on the next page which gives a little more detail about how wave functions are calculated using the Feynman 'sum over paths' approach.

3.7.2 The collapse of state

Imagine an electron moments before arriving at the double slot experiment's screen. Our best knowledge about the state of the electron comes from an array of amplitudes governing the probability that it will arrive at any individual point on the screen. One moment later the electron has been detected at one of these points. In that moment the information in the state vector collapses. One of the possibilities is made manifest—*a process that is not governed by the laws of quantum mechanics.*

Wave functions, paths and the world economy

The expression:

amplitude for an electron arriving at point x at time t

$$= \text{constant} \times \exp\left(i\left[\sum_{\text{all starting points}}\left(\sum_{\text{all paths}}\left(\frac{1}{\hbar}\int_{t_1}^{t_2} L\, \mathrm{d}t\right)\right)\right]\right)$$

can be written in a rather more compact form using mathematics. However, what is even more interesting is the manner in which the sum over paths approach can be used to relate one wave function to another. Let us say that we know the wave function of an electron $\varphi(x_a, t_a)$[12] and we are interested in finding wave function $\varphi(x_b, t_b)$ then the Feynman prescription for doing this is to calculate:

$$\varphi(x_b, t_b) = \int K[(x_b, t_b), (x_a, t_a)]\varphi(x_a, t_a)\, \mathrm{d}x_a.$$

This complex expression can be unpicked quite easily. Basically it is saying that we can get the amplitude for an electron being in place x_b at time t_b from the wave function for the electron's position at an earlier time t_a. All we have to do is to consider all the places it might have been at time t_a (that's effectively the $\varphi(x_a, t_a)$ bit) and then add it all up (that's what the integral is doing) but weighted by a function K which deals with how the electron might have travelled from the various x_a to the x_b in the time given. In other words this K function is nothing other than the sum over the various paths that we have been dealing with in this chapter. Feynman called K the *kernel*, but really it is nothing more than the amplitude for the electron to get from (x_a, t_a) to (x_b, t_b). This method for getting from one wave function to another can be generalized and is one of the real benefits of the Feynman approach. In particular, it enables physicists to calculate the effects of various fundamental forces using Feynman diagrams—which we will come across in a later chapter.

In my view, the world economy is not as interesting as quantum mechanics.

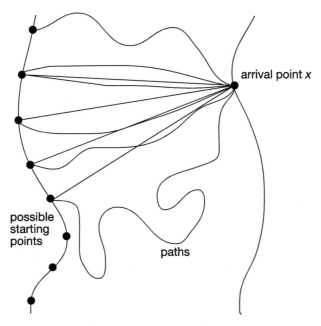

Figure 3.14. The amplitude for an electron arriving at point x, irrespective of its starting point, can be obtained by summing all the paths to x from each of a variety of starting points.

This point is extremely important. The mathematical machinery of quantum theory allows us to calculate the amplitude for a process, the state vector[13] for a particle and how both of these objects evolve with time. It would be a messy calculation, but we could do Feynman's sum over paths to calculate the probability of an electron passing through any point in the volume of space between the electron's source and the screen. This would tell us the amplitude of it passing through any point at any time. However, amplitudes or state vectors never tell us *which* point it passes through at any moment. Quantum mechanics never tells us that. The only way of discovering that is to do the experiment and find out. In the case of the double slot experiment quantum mechanics would tell us with perfect accuracy the probability of arriving at any point on the screen. With a long experimental run this would be reflected in the number of electrons arriving at one point compared to another. However, for any individual electron the mathematics of our theory does not allow us to tell which point it will arrive at. The moment the electron arrives its wave function collapses—the amplitudes for every point other than the arrival point instantly become zero and the amplitude for the arrival point becomes one.

Physicists have pondered the meaning of this for more than 80 years. In the majority of cases they have settled into a quiet life and simply gone on with calculating rather than worrying about what it means[14]. Others are not so able to forget about it and sometimes make careers out of exploring the meaning of quantum theory and the collapse of the wave function. This is a subject that is too vast to be explored in detail here, but I will make some brief remarks.

One early school of thought was that the wave function (state vector, call it what you will) represented our knowledge about the electron and the collapse was merely due to a change in our information. This is no more mysterious than the process of estimating odds on a Grand National[15] winner and then sitting back and waiting to see what happens.

Such an approach has an attraction, until one ponders the meaning of the double slot experiment for an individual electron. As we indicated earlier, if we run the experiment with only one electron crossing the apparatus at a time and then build up the pattern from individual films, the interference effect still takes place. Does this not imply that the electron has 'knowledge' of the state of the experiment (if both slits are open) that it could only have by 'sniffing out' all possible paths at once? In that case, can we say anything about where the electron is at any moment? Perhaps our ignorance of where it is going to end up is simply a reflection of the electron not knowing either! Consequently, it would be more correct to say that the electron *does not have a position*. This is a radical idea, so let me phrase it again—I am *not* saying that the electron has a position but we are in ignorance of what it is, I *am* saying that concept of position is not well defined[16] for the electron until the wave function collapses at a given place.

This issue raised its head earlier when we discussed the wave and particle way of looking at photons. On the face of it, one has difficulty in reconciling the wave aspects (interference) with the particle aspects (a single spot on the screen) of a photon or an electron. Photons and electrons are neither waves nor particles but quantum objects governed by amplitudes and state vectors. Quantum objects display wavelike behaviour when their positions are not well defined, and look more like particles when they have a well defined position. The appropriateness of the description depends on the context of the experiment. With the double slot case, the experiment is not set up to 'see' the position of the electron between its starting point and the spot on the screen where it arrives.

There is a temptation to identify the electron with the amplitude, but the amplitude describes the probability of finding an electron at a spot if we choose to look for it there. Setting up equipment to do this would change the nature of the experiment and this would also change the pattern of the amplitudes right throughout the experiment.

The most extreme view is that the amplitudes are simply the means that we use to calculate probabilities and they are not really telling us anything about the 'real' electrons at all. People who agree with this would regard electrons as convenient fictions that we have invented to relate the results of one experiment to another. They would argue that we never see an electron anyway—a spot on a photographic film is not an electron! A little pulse of current in a device at one end of the experiment and a spot on a film at the other end are the only experimentally real things. If we choose to explain this by imagining some small 'particle' called a photon moving from the device to the screen then so be it. The business of quantum theory, in this view, is simply to relate the observations together.

This is a rather extreme view, but one that does avoid a lot of problems! Most physicists have a gut feeling that the state vectors of quantum mechanics are telling us something about what electrons are. Consequently they are forced say that electrons are, in some real sense, a constantly shifting collection of possibilities with which our everyday language has no real way of coping.

If this does not take your idea of what 'real' is and give it a very good bang on the head, then I suggest you read this section again.

Quantum theory

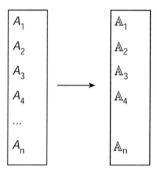

Figure 3.15. A set of amplitudes for an electron being found at positions
$1, 2, 3, 4, \ldots, n$ can be transformed into a set of amplitudes for the electron
having momentum $p_1, p_2, p_3, p_4, \ldots, p_n$.

3.7.3   The uncertainty principle

So far we have been discussing amplitudes that tell us about the
probability of an electron being found at a given place. Of course, this
is not the only sort of information we might wish to obtain about the
electron—its momentum, for example, might be of some interest.

There is an amplitude for finding the electron with any given momentum
and energy—$A(E, p)$. This object relates to the probability of finding
an electron with a given energy and momentum in an equivalent
manner to that of the amplitudes we have been treating so far. The
energy/momentum amplitude can be obtained directly with a suitable
calculation, but it can also be derived from the amplitudes for positions
and times $A(x, t)$. However, to make the transformation it is not possible
to relate a single amplitude of one type to a single amplitude of another—
the whole state vector (collection of amplitudes) has to be used (see
figure 3.15).

Figure 3.15 illustrates how a set of amplitudes for a set of fixed positions
is related to a set of amplitudes for fixed momenta. Note that in this
diagram I am not suggesting that any one amplitude on the left is directly
related to one on the right—the *set* on one side generates the *set* on the
other. In a more general case the electron is not simply restricted to
a small set of discrete positions, but can possibly be found somewhere
in a continuous span. In such a case the wave function $A(x, t)$ must

be used—a mathematical function rather than a set of values as in the diagram. The wave function can be transformed into a continuous amplitude $\mathbb{A}(E, p)$.

If the amplitude for finding an electron is fairly localized in space (e.g. that $A(x_3, t)$ and $A(x_4, t)$ are large and the others are very small) the mathematical machinery generates a set of $\mathbb{A}$ amplitudes that are very similar in size for a range of p values. It turns out that the only way of producing a set of $\mathbb{A}$ amplitudes that are large only for p values very close to a given value, is to work from a set of A amplitudes that give the electron a good chance of being anywhere! This discovery was first made by the physicist Werner Heisenberg and is now known as *Heisenberg's uncertainty principle* and can be summarized in the relationship:

➡ $\Delta x \times \Delta p \geqslant h/2\pi$, where h is the ubiquitous Planck's constant.

In strict mathematical terms what this is telling us is as follows:

> Δx is the range of x values (positions) about a given value of x for which the amplitude A is large;
> Δp is the range of p values (momenta) about a given value of p for which the amplitude $\mathbb{A}$ is large.

If you multiply these two ranges together, then the answer cannot be less than $h/2\pi$.

Like all matters to do with quantum theory, one needs to be very careful about the meaning of this principle. The name doesn't help the search for clarity[17]. It would be very easy to interpret the *uncertainty* principle as a lack of knowledge about the position or momentum of the electron. It would be more true to say that the principle explains the extent to which our concepts of position or momentum can be applied to a given electron.

If we set up an experiment that makes the position amplitude fairly localized in space, then we can only do this at the cost of saddling the electron with an amplitude that is significant for a range of momentum values. Under these conditions it would be better to say that the electron does not have a well defined momentum. On the other hand, if we carry out an experiment in which the electron's momentum *is* well defined (i.e. $\mathbb{A}$ is only large for a few values of p close to a specified value[18]) then

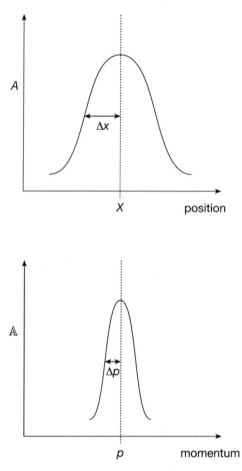

Figure 3.16. The uncertainty principle relates a range of amplitudes for position
(*A*) to a range of amplitudes for momentum (𝔸). The smaller Δ*x* is the greater
Δ*p* must be and vice versa.

the electron's position loses much meaning (i.e. at any moment in time
it has a significant amplitude to be in a wide range of places).

Position and momentum are not the only variables to be linked by an
uncertainty principle of this form. Another example would be energy

and time:

➡ $$\Delta E \times \Delta t \geqslant h/2\pi$$

an expression that needs even more careful interpretation than the position/momentum uncertainty. The issue here is the ambiguity in what is meant by t—it can be either a *moment* in time (right now it is 5.05pm) or *duration* of time (this train journey has taken 2 hours[19]). Often the distinction between these is blurred when it should be made clear.

The uncertainty principle is really involved with durations of time. As always with quantum theory the experimental set-up involved can be a guide to the meaning.

If we allow an electron to interact with some other object (a measuring apparatus for example), the interaction will last for a certain duration Δt. If this duration happens to be very short, then the electron's interaction will involve amplitudes for it to have an energy value lying within a certain (wide) range ΔE of a specific value E. On the other hand, if the interaction is for a long duration (large Δt), then this will force the amplitudes for the various E values to be very small unless they are close to the specific value E (i.e. ΔE is small).

The really remarkable thought that has emerged over the last ten years or so is that the uncertainty principle might be directly responsible for the formation of galaxies. If our current ideas are correct, then early in the universe's history it underwent a period of dramatic expansion governed by a quantum field. The energy/time uncertainty principle was reflected in parts of the field which have slightly more energy than other parts. This was eventually converted into parts of the universe being slightly more dense than other parts—eventually these density fluctuations gave rise to the large-scale structures that we see now.

There is even the suggestion that the universe as a whole may be a fluctuation governed by uncertainty...

These are very dramatic and exciting new ideas that we will take up more fully in the closing chapters of the book.

3.8   Summary of chapter 3

- The ordinary rules of probability do not apply in the microscopic world of atoms and subatomic particles;
- quantum theory is a new way of calculating that was developed to cope with the strange results of atomic experiments;
- in quantum mechanics every event has an amplitude which is squared to give the probability of the event happening;
- amplitudes can be added and multiplied to account for alternative 'paths' to the event;
- amplitudes sometimes combine to increase the overall amplitude and sometimes they combine to decrease the overall amplitude;
- Feynman's rules for calculating amplitudes state that every path to an event must be considered, no matter how far it is from the common sense 'classical' path;
- the key to calculating the amplitude of a path is the Lagrangian of the particle moving along that path;
- the 'real' world is the quantum world; what we see in our large-scale world is an average of the quantum effects that are actually going on all the time;
- the wave function is a collection of amplitudes covering a range of different positions;
- given the wave function at one place and time the wave function at another place and time can be calculated by connected to two using Feynman's paths;
- amplitudes for position and time are related to amplitudes for energy and momentum;
- Heisenberg's uncertainly principle related position/momentum and energy/time together so that if the amplitude for one is tightly focused on a given value, the amplitude for the other is spread over a range of values.

Notes

[1] Einstein said that common sense was a prejudice laid down before the age of 18.

[2] Strictly speaking, the single slot pattern is different to that one might get from ricocheting bullets. When quantum effects are taken into account there is also a diffraction pattern from a single slit. However, if the width of the slit is very

much greater than the wavelength of the electrons (see later) then the pattern is rather similar to that one might expect from bullets.

[3] Electron double slit experiments are not easy to do! The first such experiment was performed by G Mollenstedt and H Duker in 1954. These days they are done with beams from an electron microscope which typically have energies of 60 keV. The beam is aimed at a pair of slits about 0.5 μm wide separated by 2 μm. With a detector placed about 2 m from the slits the distance between bright lines on the pattern is 5 μm.

[4] Although I have not suggested any mechanism that might enable this 'communication' between electrons, it is important to eliminate it as a possibility.

[5] Undergraduates learn the 'older' view of quantum theory first as they need to develop their mathematical skills before they can go on to Feynman's view. Often they will not see the later way of doing calculations unless they go on to graduate work in the subject.

[6] Apply this to the UK lottery. Each lottery ball has 49 numbers and 6 numbers need to be correct. The chances of selecting the right sequence are

$$\frac{6 \times 5 \times 4 \times 3 \times 2 \times 1}{49 \times 48 \times 47 \times 46 \times 45 \times 44}$$

i.e. 1 in 13 983 816. Not good odds!

[7] Barring such effects as the difference in mass and size etc, which can be compensated for in the calculations.

[8] One sometimes hears quantum objects talked about as a set of connections and relationships, rather than as a set of fixed properties.

[9] The task of finding an appropriate Lagrangian is central to discovering the laws of nature that operate in a given situation. There are general rules that can be used to help guess the right Lagrangian (it must relate well to relativity, conserve electrical charge and energy etc) but in the end, until a deeper understanding is found, the Lagrangian must be produced by educated guesswork and then tested against experimental results.

[10] Consider two amplitudes—one that takes the electron round an orbit and back in a given time, and another that allows the electron to just sit there for the same amount of time. In both cases the final result is the same—so should the amplitudes for the final state not be the same?

[11] It is also often written as $\phi(x, t)$ or $\psi(x, t)$.

[12] Here I have switched to using the Greek letters for the wave function, as that is the symbolization more generally used in quantum mechanics books.

[13] Remember that amplitudes tell us about one possibility, the state vector contains the amplitudes for all the possibilities.

[14] Bills have to be paid after all.

[15] For those that do not know, this is a big horse race in England.

[16] By this I mean that there is no single position that we can accuse the electron of being in at any one moment.

[17] Neither do some of the explanations found in various popular books.

[18] It is virtually impossible to set up an experiment that produces precisely one momentum.

[19] Guess where I am writing this section. . .

Chapter 4

The leptons

In this chapter we shall study the leptons and introduce the way in which we describe particle reactions. In the process we shall learn how conservation of electrical charge applies to particle physics. The solar neutrino problem will be discussed and some features of the weak force will become apparent.

4.1 A spotter's guide to the leptons

Leptons, unlike quarks, can exist as separate objects (remember that the strong force binds quarks into particles, so that we can never study a single quark without its partners). Hence their reactions are more straightforward than those of the quarks. This makes it easier to develop some of the ideas of particle physics using leptons as examples. Later we can apply these ideas to the slightly more complicated case of the quarks.

Particle physics can seem like a very 'unreal' subject which studies rare and unimportant sounding objects. This is not the case. Table 4.1 shows that some of the leptons are readily found in nature and have important roles to play in common processes.

The electron is a well-known particle and its properties are established in basic physics courses. Its partner in the first generation, the electron-neutrino, is less well known but just as common in nature. Some radioactive processes produce them and the central cores of atomic

85

Table 4.1. Where to find leptons.

1st generation	2nd generation	3rd generation
electron	muon	tau
(i) found in atoms (ii) important in electrical currents (iii) produced in beta radioactivity	(i) produced in large numbers in the upper atmosphere by cosmic rays	(i) so far only seen in labs
electron-neutrino	muon-neutrino	tau-neutrino
(i) produced in beta radioactivity (ii) produced in large numbers by atomic reactors (iii) produced in huge numbers by nuclear reactions in the sun	(i) produced by atomic reactors (ii) produced in upper atmosphere by cosmic rays	(i) so far only seen in labs

reactors emit them in very large numbers. The sun is an even more copious producer of electron-neutrinos. Approximately 10^{12} electron-neutrinos pass through your body every second, the vast majority of which were produced by the nuclear reactions taking place in the sun's core. Fortunately, neutrinos interact with matter so rarely that this huge number passing through us does no harm whatsoever.

The members of the second generation are less common, but they are still seen frequently enough in nature to make their study relatively straightforward.

Muons are easily produced in laboratory experiments. They are similar to electrons, but because they are more massive they are unstable and decay into electrons and neutrinos (in just the way an unstable atom will decay radioactively). Unlike electrons they are not produced by radioactive decay. They are easily observed in cosmic ray experiments.

Table 4.2. The leptons (M_p = proton mass). Charge of particles is -1 in the top row and 0 in the bottom row.

1st generation	2nd generation	3rd generation
electron	muon	tau
$5.45 \times 10^{-4} M_p$ $(5.11 \times 10^{-4} \text{ GeV}/c^2)$	$0.113 M_p$ $(0.106 \text{ GeV}/c^2)$	$1.90 M_p$ $(1.78 \text{ GeV}/c^2)$
electron-neutrino	muon-neutrino	tau-neutrino
$2\text{--}3 \times 10^{-9} M_p$ $(\sim 2\text{--}3 \text{ eV}/c^2)$	$< 1.8 \times 10^{-4} M_p$ $(< 170 \text{ heV}/c^2)$	$1.9 \times 10^{-2} M_p$ $(< 18.2 \text{ MeV}/c^2)$

The members of the third generation have not been seen in any naturally occurring processes—at least not in the current era of the universe's evolution. In much earlier times, when the universe was hotter and particles had far more energy, the members of the third lepton generation would have been produced frequently in particle reactions.

This, however, was several billion years ago. Today we only observe the tau in experiments conducted by particle physicists. The tau-neutrino has not been observed directly, but its presence can be inferred from certain reactions.

4.2 The physical properties of the leptons

Table 4.2 reproduces the standard generation representation of the leptons.

The electron, muon and tau all have an electrical charge of -1 on our standard scale, while the three neutrinos do not have an electrical charge.

Table 4.2 also shows that the charged leptons increase in mass from generation to generation. However, the neutrino masses are not so clear cut.

Our best experimental measurements have been unable to establish a definite mass for any of the neutrinos. The numbers in table 4.2 are the upper limits that we are able to place on the various neutrino masses. The increase in these numbers across the table displays our increasing *ignorance* of the masses, not necessarily an increase in the *actual* masses of the neutrinos. It is not possible to 'capture' just one example of a particle and place it on a set of scales. Masses have to be calculated by studying the ways in which the particles react. The more particles one measures, the more precise one can be about the size of their mass.

For example, the tau-neutrino is a very rare object that is difficult to study. All we can say at this time is that its mass (if it has one) is less than 0.019 proton masses. This limit comes from studying the decay of the tau, which is itself a rare object. The uncertainty will become smaller as we measure more of them. The electron-neutrino, on the other hand, is much more common, so the upper limit on its mass is much more clearly defined.

Unfortunately, our theories do not predict the masses of the neutrinos. For a long time particle physicists assumed that neutrinos have no mass at all, but recently some measurements have suggested that they may have an extremely small, but finite, mass. This is an open question in experimental physics. The 'standard version' of the standard model assumes that the neutrinos have no mass. If they do, then the theories can be adapted to include this possibility.

Neutrinos seem like very weird particles. They may have no mass (although professional opinion is swinging towards their having *some* mass) and they certainly have no charge, so they are very ghostly objects. Being leptons they do not feel the strong force and having no electrical charge they do not feel the electromagnetic force either. The only way that neutrinos can interact is via the weak force. (Ignoring gravity, which would be extremely weak for an individual neutrino. However, if there are enough neutrinos with mass in the universe, their *combined* gravity might stop the expansion of the universe.)

As weak force reactions are very rare, a particle that *only* reacts by the weak force cannot be expected to do very much. Estimates indicate that a block of lead *ninety thousand million million* metres thick would be required to significantly reduce the number of electron-neutrinos passing through your body at this moment.

4.3 Neutrino reactions with matter

Neutrinos are difficult to use in experiments, but they are worth the effort for one reason: they are the only particles that allow a relatively uncluttered study of the weak force. After all, they do not react in any other way. Their reactions may be very rare, but if you have enough neutrinos to begin with then sufficient reactions can take place for a sensible study to be made.

A characteristic reaction involving a neutrino is symbolized in equation (4.1):

$$\nu_e + n \rightarrow p + e^-. \tag{4.1}$$

This reaction has been of considerable experimental importance over the past few years. Astrophysicists have calculated the rate at which electron-neutrinos should be being produced by the sun. In order to check this theory physicists have, in several experiments, measured the number of neutrinos passing through the earth. The first of these experiments relied on the neutrino–neutron reaction (4.1).

If the neutron in (4.1) is part of the nucleus of an atom, then the reaction will transform the nucleus. For example, a chlorine nucleus containing 17 protons would have one of its neutrons turned into a proton by the reaction, changing the number of protons to 18 and the nucleus to that of argon:

$$^{37}_{17}\text{Cl} + \nu_e \rightarrow {}^{37}_{18}\text{Ar} + e^-. \tag{4.2}$$

The electron produced by the reaction would be moving too quickly to be captured by the argon nucleus. It would escape into the surrounding material. However, the transformation of a chlorine nucleus into an argon nucleus can be detected, even if it only takes place very rarely. All one needs is a sufficiently large number of chlorine atoms to start with.

The Brookhaven National Laboratory ran an experiment using 380 000 litres of chlorine-rich cleaning fluid in a large tank buried in a gold mine many metres below the ground. Few particles can penetrate so far underground and enter the tank. Mostly these will be the weakly reacting solar neutrinos. To isolate the equipment from particles produced in the rock, the tank was shielded. The energy of solar neutrinos is sufficient to convert a nucleus of the stable isotope chlorine 37 into a nucleus of argon 37. Argon 37 is radioactive. Every few months the tank was flushed out and the argon atoms chemically separated from the chlorine. The number of argon atoms was then counted by measuring the intensity of the radiation that they produced.

Theoretical calculations show that the number of reactions should be approximately 8 per second for every 10^{36} chlorine 37 atoms present in the tank. The experiment ran from 1968 to 1986 and over that time measured only 2 per second for every 10^{36} atoms in the tank. Both numbers represent incredibly small rates of reactions (a measure of how weak the weak force is) but the conclusion is clear—the measured number of electron-neutrinos produced from the sun is totally different from the theoretical predictions. This is the solar neutrino problem.

This is an example of a very concrete piece of physics based on the properties of these ghostly particles, and a genuine mystery in current research.

4.3.1 Aspects of the neutrino–neutrino reaction

4.3.1.1 *Conservation of electrical charge*

Returning to the basic reaction (4.1) within the chlorine nucleus:

$$\nu_e + n \rightarrow p + e^- \tag{4.1}$$

the only way that a neutron can be turned into a proton is for one of the d quarks to be turned into a u. Evidently, inside the neutron, the following reaction has taken place:

$$\nu_e + d \rightarrow u + e^-. \tag{4.3}$$

This is the basic reaction that has been triggered by the weak force. Notice that it only involves one of the quarks within the original neutron. The other two quarks are unaffected. This is due to the short range of the weak force. The neutrino has to enter the neutron and pass close to one of the quarks within it for the reaction to be triggered.

This is quite a dramatic transformation. The up and down quarks are different particles with different electrical charges. Examining the reaction more carefully:

$$\nu_e \; + \; d \; \to \; u \; + \; e^-$$

charges $\quad 0 \qquad\quad -\frac{1}{3} \qquad +\frac{2}{3} \qquad -1$

we see that the individual electrical charges of the particles involved have changed considerably. However, if we look at the *total* charge involved then we see that the total is the same after the reaction as before. On the left-hand side the charge to start with was $(0) + (-1/3) = -1/3$, and on the right the final total charge of the two particles is $(+2/3) + (-1) = -1/3$ again.

This is an example of an important rule in particle physics:

CONSERVATION OF ELECTRICAL CHARGE

In any reaction the total charge of all the particles entering the reaction must be the same as the total charge of all the particles after the reaction.

Although we have illustrated this rule by the example of a weak force reaction involving neutrinos the rule is much more general than that. Any reaction that can take place in nature must follow this rule, no matter what force is responsible for the reaction. It is good practice to check the rule in every reaction that we study.

4.3.1.2 *Muon neutrinos and electron-neutrinos*

There is an equivalent reaction to that of the electron-neutrino involving muon-neutrinos:

$$\nu_\mu + n \to p + \mu^-. \tag{4.4}$$

In this case the neutrino has struck the neutron transforming it into a proton with the emission of a muon. The incoming muon-neutrino has to have more energy than the electron-neutrino as the muon is 200 times more massive than the electron. The muon will be produced with a large velocity and will be lost in the surrounding material. The reaction is very similar to the previous one. At the quark level:

$$\nu_\mu + d \to u + \mu^- \tag{4.5}$$

a reaction that also conserves electrical charge.

There is one important contrast between the reactions. In all the reactions of this sort that have ever been observed, the muon-neutrino has *never* produced an electron, and the electron-neutrino has never produced a muon[1].

This is a prime example of the weak force acting *within* the lepton generations. The electron and its neutrino are in the first lepton generation; the muon and its neutrino are in the second generation. The weak force cannot cross the generations. It was the contrast between these two reactions that first enabled physicists to be sure that there were different types of neutrino.

The existence of the neutrino was predicted by the Viennese physicist Wolfgang Pauli[2] in 1932. Unfortunately, it was not until atomic reactors were available in the 1950s that sufficient neutrinos could be generated to allow the existence of the particles to be confirmed experimentally. The large numbers produced by a reactor core were needed to have sufficient reactions to detect[3].

By the late 1950s the existence of the muon had been clearly established (it had first shown up in cosmic ray experiments in 1929) and its similarity to the electron suggested that it might also have an associated neutrino. However, as neutrino experiments are so hard to do, nobody was able to tell if the muon-neutrino and the electron-neutrino were different particles.

The problem was solved by a series of experiments at Brookhaven in the 1960s. A team of physicists, lead by Leon Lederman, managed to produce a beam of muon-neutrinos[4]. The neutrino beam was allowed to interact with some target material. Over a 25-day period some 10^{14} muon-neutrinos passed through the experiment. The reactions between the neutrinos and the nuclei in the target produced 51 muons. No electrons were produced. In contrast, electron-neutrinos readily produced electrons when passed through the target material, but no muons. The results of Lederman's experiment were:

$$\nu_\mu + n \rightarrow p + \mu^- \qquad \text{51 events}$$
$$\nu_\mu + n \rightarrow p + e^- \qquad \text{no events}$$

but

$$\nu_e + n \rightarrow p + e^-$$

so the v_e and the v_μ are different! This experiment clearly established the fundamental distinction between these two types of neutrino.

A later series of experiments at CERN found the ratio of electrons to muons produced by muon-neutrinos to be 0.017 ± 0.05, consistent with zero. For his pioneering work in identifying the two types of neutrino Lederman was awarded the Nobel Prize in physics in 1989[5].

Now, of course, we realize that there are three neutrinos, the tau-neutrino being in the third generation of leptons along with the tau.

4.4 Some more reactions involving neutrinos

The fundamental distinction between muon-neutrinos (v_μ) and the electron-neutrinos (v_e) is reflected in a variety of reactions. For example, there is an interesting variation on the basic reaction that lies behind the solar neutrino counting experiment.

The basic v_e reaction converts nuclei of chlorine 37 into radioactive argon 37. It is also possible for the argon 37 atom to reverse this process by 'swallowing' one of its electrons:

$$^{37}_{18}\text{Ar} + \text{e}^- \rightarrow {}^{37}_{17}\text{Cl} + v_e. \tag{4.6}$$

This reaction is called 'K capture' because the electron that triggers the reaction comes from the lowest energy state of the orbital electrons, called the K shell. Electrons in this lowest energy state spend much of their time in orbit very close to the surface of the nucleus. Sometimes they pass *through* the outer layers of the nucleus. With certain atoms, and argon 37 is an example, this electron can be 'swallowed' by one of the protons in the nucleus turning it into a neutron. This process is another example of a reaction triggered by the weak force. Once again the weak force must work within the lepton generations, so the member of the first generation that was present before the reaction, the electron, has to be replaced by another member of the first generation after the reaction—in this case the v_e that is emitted from the nucleus. At the underlying quark level:

$$
\begin{array}{ccccccc}
\text{u} & + & \text{e}^- & \rightarrow & \text{d} & + & v_e \\
\tfrac{2}{3} & + & -1 & = & -\tfrac{1}{3} & + & 0.
\end{array}
\tag{4.7}
$$

Again we note that charge is conserved in this reaction.

Another reaction that illustrates the weak force staying within generations is the collision of a v_μ with an e^-:

$$v_\mu + e^- \rightarrow \mu^- + v_e. \tag{4.8}$$

This is a very difficult reaction to study. Controlling the neutrino beam and making it collide with a sample of electrons is hard to arrange.

In this reaction a v_μ has been converted into a μ^-, and an e^- has been converted into a v_e. Notice that the weak force has again preserved the generation content in the reaction. The particles have changed, but there is still one from the first generation and one from the second on each side of the reaction.

We have now accumulated enough evidence to suggest the following rule:

LEPTON CONSERVATION
In any reaction, the total number of particles from each lepton generation must be the same before the reaction as after.

All the reactions that we have looked at have followed this rule (go back and check). Physicists regard this as one of the fundamental rules of particle physics. There are two points worth noting about this rule:

- it must also be followed by the other forces, as the weak force is the only one that allows fundamental particles to change type;
- it is not clear how the rule applies if there are no particles from a given generation present initially.

The second point can be illustrated by an example from nuclear physics.

Beta radioactivity takes place when a neutron inside an unstable isotope is transformed into a proton by the emission of an electron. In terms of observed particles the reaction is:

$$n \rightarrow p + e^-. \tag{4.9}$$

The reaction is incomplete as it stands. When β decay was first studied only the electron was observed, which was a puzzle as the reaction did not seem to conserve energy or momentum (see section 8.2.4). We can also see that it does not appear to observe the lepton conservation rule.

At the quark level there is a fundamental reaction taking place:

$$d \rightarrow u + e^-. \tag{4.10}$$

As a quark has changed type, we must suspect that the weak force is involved[6]. However, on the left-hand side of the equation there is no member of the first generation present. On the right-hand side an electron has appeared—so now there is a first generation representation. This is an odd situation. Either the rule is violated in certain circumstances, or we have not yet found out how to consistently apply the rule. In the next chapter we shall explore this in more detail and discover how we can make the rule *always* apply.

4.5 'Who ordered that?'[7]

So far we have concentrated on the massless (or at least very low mass) particles in the lepton family. We have not really discussed the reactions and properties of the massive leptons. In part this is because in broad terms they behave exactly like electrons with a greater mass.

For example, muons are common particles that are easily produced in reactions. They have a negative charge which all measurements indicate is *exactly* the same as that of the electron. The only apparent difference between muons and electrons is that the muon is 200 times more massive. We have no idea what makes the muon so massive.

The muon was first discovered in experiments on cosmic rays[8]. Cosmic rays are a stream of particles that hit the earth from the sun and other stars in the galaxy. 75% of the cosmic rays counted at sea level are muons. These muons, however, do not originate in outer space. They are produced high up in the earth's atmosphere. Protons from the sun strike atoms in the atmosphere causing reactions that produce muons[9]. We can be quite sure that the muons we observe at sea level have not come from outer space because the muon is an unstable particle.

It is impossible to predict when a given muon will decay (just as it is impossible to predict when a given atom will decay), but we can say on

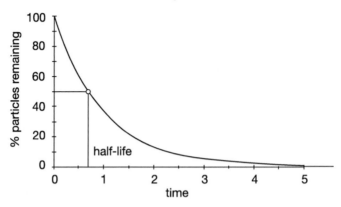

Figure 4.1. The decay of particles.

average how long a muon will live for. Particle physicists call this the *lifetime* of the particle.

Figure 4.1 shows how the number of particles decreases with time in a batch of identical particles that were all created at the same moment. The same shape of curve would be produced if the number of radioactive atoms were plotted instead. The half-life is the amount of time taken to reduce the number of particles (or atoms) by 50%. The lifetime is the average life of a particle in the set.

Muon lifetimes have been measured very accurately; the quoted value is $(2.197\,03 \pm 0.0004) \times 10^{-6}$ seconds (i.e. about 2 millionths of a second[10]). This seems a remarkably short length of time, but compared to some the muon is a very long lived particle. This is why the muons *must* be produced in the upper atmosphere. They do not live long enough to reach the surface of the earth from any further away. Indeed, most muons produced in the upper atmosphere decay before they reach sea level. That any muons manage to survive is a direct consequence of Einstein's theory of relativity which predicts that time passes more slowly for moving objects[11].

The physics of decay processes is worthy of a chapter on its own (chapter 8) but we can briefly discuss one important point here: the particles produced by the decay *did not exist before the decay took place*.

This is quite difficult to imagine. The muon, for example, does not consist of an electron and some neutrinos bound together. As we have been saying for some time, the muon is a fundamental particle and so does not have any objects inside it. The decay process is not like a 'bag' splitting open to release the electron and neutrinos that were held inside. At the moment of decay the muon disappears and is replaced by the electron and the neutrinos. This seemingly magical process is fully explained by Einstein's theory of relativity.

If you have difficulty believing that such a process can take place, then consider the emission of light by electrons. Atoms emit light when one of their electrons drops from a high energy level to a lower level. Ask yourself if the light wave (or photon) existed *before* the electron dropped levels. Would it be true to say that the electron was composed of an electron and a light wave before the emission?

Decay processes and the emission of light are both examples of the transformation of energy from one type to another. The electron loses energy as it drops a level and this energy is emitted in the form of a light wave (or photon). Similarly, the energy locked up in the muon is turned into a different form—the electron and neutrinos. This turning of energy into matter is a consequence of $E = mc^2$.

The tau is also unstable. It can decay, for example, into a muon and neutrinos (the muon will in turn decay as well). Being more massive than the muon (about 17 times) the tau has an even smaller lifetime of $(3.3 \pm 0.4) \times 10^{-13}$ seconds. The tau is not as well studied as the other leptons. It is a comparatively recent discovery (1975) and difficult to produce in large numbers.

The decays of the muon and tau raise another point about the lepton conservation rule:

$$\mu^- \to e^- + \nu_\mu + \nu_? \tag{4.11}$$

$$\tau^- \to \mu^- + \nu_\tau + \nu_?. \tag{4.12}$$

We can deduce the presence of the ν_μ in (4.11) from the rule of lepton conservation. We started with a member of the second generation (μ^-), so we must end up with a member of the second generation (ν_μ). The presence of a second neutrino among the decay products is not so obvious. How can we be sure that it is there?

The existence of the second neutrino can be deduced by measuring the energy of the electrons. The result shows that the electrons vary considerably in energy from one decay to another. If there was only one neutrino, then there would only be one way of sharing the energy out between the two particles. Hence the electrons would always have the same energy. If there are three particles, then there are an infinite number of ways of sharing the energy among the three—hence we can find the electron with a spread of energies[12].

We have argued that one of these neutrinos must be a ν_μ on the grounds of lepton conservation, but what of the other? As it stands the reaction does not conserve leptons overall. The number of second generation particles balances, but there is a first generation particle among the decay products (the e^-) and there was no such particle initially.

This is the same problem that we saw in the context of neutron decay. The problem is repeated in the tau decay equation (4.12). The rule does not seem to work as we understand it at the moment. We need to develop a more complete understanding in the next chapter.

4.6   Solar neutrinos again

In June 2001 the first results were announced from an experiment that was designed to help solve the solar neutrino problem. The Sudbury Neutrino Observatory (SNO)[13] was built over 2 km under ground, in INCO's Creighton mine near Sudbury, Ontario. The detector uses 1000 tonnes of heavy water in a 12 m diameter acrylic vessel. Neutrinos react with the heavy water to produce flashes of light called Čerenkov radiation (see section 10.4.3), which can be electronically detected. The key aspect of this experiment is that it can detect all three types of neutrino. One solution to the solar neutrino problem is that the sun is producing the correct number, but some of them are transformed into different neutrinos on the way to the earth. Previous experiments have only been able to detect electron-neutrinos. The analysis of the first set of data from the experiment seems to show impressive confirmation that these 'neutrino oscillations' are taking place—if the fluxes from all types of neutrino detected are collected together, then a value consistent with that predicted from standard solar physics is measured. If neutrino oscillations are taking place, then the neutrinos themselves must have small masses. The currently available data on this suggests that the electron-neutrino has a mass <2.8 eV and that the sum of the masses of all three neutrinos must lie between 0.05 and 0.18 eV.

4.7 Summary of chapter 4

- The electron, the muon and their neutrinos are very common fundamental particles in the universe;
- the tau is too heavy to be produced naturally at this stage of the universe's evolution, but was much more common billions of years ago;
- the tau-neutrino is not common today because the tau is not produced;
- all three neutrinos may have very small masses;
- neutrinos can react with matter, but very rarely, so experiments have to use large numbers of them;
- experiments to measure the number of electron-neutrinos from the sun are currently finding far fewer than theory would suggest;
- *all reactions follow the rule of conservation of electrical charge*;
- the weak interaction only works within lepton generations.

Notes

[1] Even if the electron-neutrino had sufficient energy to create the muon.

[2] Pauli is well known for the Pauli principle—that electrons in atoms cannot sit in exactly the same state as each other—and the Pauli effect—that any equipment would spontaneously go wrong in his presence. Pauli was a *theoretical* physicist.

[3] The experiment was performed by a team lead by F Reines between 1953 and 1956. By the end of this time they were able to confirm that of the 10^{20} neutrinos produced by the reactor every second, one reacted with their equipment every twenty minutes. Pauli died in 1958, his prediction confirmed after a 24-year wait.

[4] They used the decay $\pi^+ \rightarrow \mu^+ + \nu_\mu$ and filtered out the muons by passing them through lead sheets.

[5] There is a well known story regarding Lederman's experiment. The team was using iron from a decommissioned battleship to filter out other particles. This involved packing the grooves inside one of the battleship's gun barrels with wire

wool in order to produce a uniform volume of material. After an especially hard day crawling along the gun barrel, a rather upset graduate student stormed into Lederman's office to resign. 'You are never going to get me down that gun barrel again' he is supposed to have said. Lederman's reply is worthy of a Nobel Prize in itself: 'You must go back, we have no other students of your calibre!'

[6] You may be wondering if we can really say that a force is involved as there is only one particle on the left-hand side—how can a force act on one particle? This is a very deep subject in particle physics we tackle in chapter 8.

[7] A remark attributed to the theorist Isidore Rabi on hearing of the existence of the muon and its similarity to the electron.

[8] Carl Anderson and Seth Neddermeyer first observed muon tracks in the early 1930s.

[9] Strictly, the reactions produce particles called pions which then decay into muons; see chapter 8.

[10] Lifetimes of this size are characteristic of the involvement of the weak force.

[11] This very odd sounding prediction has been confirmed by flying an atomic clock in a high-speed plane. Before the flight it was synchronized with a duplicate clock and when compared again after the flight was found to have lost time.

[12] This is quite a tricky argument involving conservation of energy and momentum. We will look at it again in more detail when we discuss neutron decay in chapter 8.

[13] Further information on this experiment and related work can be found on the web sites listed in appendix 4.

Chapter 5

Antimatter

In this chapter we shall examine the generation structure of weak force reactions. Lepton numbers are introduced as internal properties and extended to cater for antileptons. We shall anticipate chapter 6 by introducing baryon number and the antiquarks. Some general, but important, remarks will be made about the nature of antimatter. Finally, we shall consider lepton/antilepton reactions.

5.1 Internal properties

The weak interaction can change a lepton into its generation partner, but this conversion cannot take place between different generations. The weak force also acts predominantly within quark generations, but it is able to cross generations with a reduced effect. The conservation of leptons rule is based on this information. However, the rule seems to get into trouble when applied to the decay of particles such as:

$$\mu^- \to e^- + \nu_\mu + \nu_? \qquad (4.11)$$
$$\tau^- \to \mu^- + \nu_\tau + \nu_?. \qquad (4.12)$$

One of the produced neutrinos must be of the same generation as the original particle to conserve the number of particles in that generation. The second neutrino, however, is a complete mystery. The solution to this puzzle reveals what it is that makes the lepton generations so distinct.

Table 5.1. Assignments of lepton numbers.

	Electron-number L_e	Muon-number L_μ	Tau-number L_τ
electron	1	0	0
electron-neutrino	1	0	0
muon	0	1	0
muon-neutrino	0	1	0
tau	0	0	1
tau-neutrino	0	0	1

The electron, muon and tau are very similar particles, the only difference between them being mass. On its own, this difference might be enough to separate them into distinct generations. However, the situation is far less clear for the three neutrinos. There is no compelling evidence, at the moment, that any of them have any mass at all, never mind different masses. If that is the case, then we are hard pressed to say what the physical difference between them is! They are all zero mass, zero charge objects. Yet Lederman's work convinced us of the difference between the ν_e and the ν_μ. There must be some property that distinguishes them, even if we cannot measure it.

Let us consider the possibility that there is some *internal* property that we cannot measure by any conventional means (such as we might use to measure charge and mass) which distinguishes the generations of leptons. We call this property *lepton number* and it splits into three different 'values' L_e, L_μ and L_τ. Table 5.1 shows how this new property is assigned to the various leptons.

Imagine that there is a 'switch' that sits inside every lepton. This switch can be set in one of three ways that determines the lepton generation that the particle belongs to. To record the position of the switch we need three numbers L_e, L_μ and L_τ. These numbers can be either 1 or 0. A member of the first generation has $L_e = 1$ but $L_\mu = 0$ and $L_\tau = 0$. Of course we do not need to keep on saying which numbers are zero, it is sufficient to identify the number that is equal to 1.

Notice that the switch does not distinguish between the lepton and its neutrino: that is done by mass. Lepton number only distinguishes between generations. Experiments cannot read the value of this switch directly, but we can tell how it is set by observing the particle's reactions with other particles. Of course this switch does not actually exist—there are no mechanisms inside a particle—this is just a way of thinking about lepton number.

One reason why we cannot measure such a property is that it does not have a size, in the sense that charge or mass does. Electron-number cannot be 1.4 or 2.7 or 1.6×10^{-19} or any other number. A particle has either got electron-number or it hasn't—they are the only two possibilities. Physicists call such properties of particles (and we shall come across others) *internal properties*[1].

So far all we have done is to 'invent' a series of numbers that separate the various lepton generations. A physicist would need to be convinced that we are doing something more interesting than just playing with numbers. We need to ensure that lepton number is a real physical property.

There are two basic conditions that, if satisfied, will convince most physicists that a given property is real and not an invention of our imagination:

- it can be demonstrated that lepton number is a conserved quantity in particle reactions;
- using the lepton number conservation rule helps us to understand reactions that we could not understand or predict otherwise.

Happily, both of these conditions are satisfied.

5.1.1 Lepton number conservation

Consider a reaction from the previous chapter:

$$\nu_e + d \rightarrow u + e^-. \tag{4.3}$$

By definition all quarks have lepton number $= 0$ (they are not leptons!). The ν_e and the electron both have $L_e = 1$. Hence the total lepton numbers of the particles before and after the reaction are:

$$
\begin{array}{ccccccc}
& \nu_e & + & d & \rightarrow & u & + & e^- \\
L_e & 1 & + & 0 & = & 0 & + & 1.
\end{array}
$$

Evidently L_e is conserved in this reaction. We need not consider L_μ and L_τ as the only leptons in the reaction are from the first generation.

Taking another example:

$$\nu_\mu + e^- \rightarrow \mu^- + \nu_e. \tag{4.8}$$

In this case the total electron-number (L_e) and the total muon-number (L_μ) *separately* must be the same before and after the reaction:

$$
\begin{array}{ccccccc}
 & \nu_\mu & + & e^- & \rightarrow & \mu^- & + & \nu_e \\
L_e & 0 & + & 1 & = & 0 & + & 1 \\
L_\mu & 1 & + & 0 & = & 1 & + & 0.
\end{array}
$$

We could consider many other reactions that would all demonstrate the conservation of the various lepton numbers. This has become a well established rule in particle physics amply confirmed by experiment and with a secure theoretical grounding as well.

LEPTON NUMBER CONSERVATION
The total electron-number, muon-number and tau-number are
separately conserved in all reactions

Note the similarity between this rule and the conservation of electrical charge.

By this stage you may have the uncomfortable feeling that all we are doing is describing what is essentially a very simple thing in a complex way. In the last chapter we simply said 'the weak force must maintain the number of leptons of each generation that are involved in a reaction'. Now we are conserving three different lepton numbers.

This would be a serious criticism if it could not be shown that lepton number conservation leads us into new physics that we could not otherwise explain.

5.1.2 Mystery neutrinos

There are some reactions that seem to violate the rule that we now call lepton number conservation. One of the reactions that is giving us trouble is:

$$\mu^- \to e^- + \nu_\mu + \nu_?. \tag{4.11}$$

If we look at this reaction in terms of lepton number then something interesting emerges:

$$
\begin{array}{ccccccc}
 & \mu^- & \to & e^- & + & \nu_\mu & + & \nu_? \\
L_\mu & 1 & = & 0 & + & 1 & + & ? \\
L_e & 0 & = & 1 & + & 0 & + & ?
\end{array}
$$

The assignment of lepton number for the mystery neutrino is unclear as we have not identified what sort of object this is. One thing seems very likely from this however—*it must have muon-number = 0*. If this were not the case, the muon-number conservation law would certainly be violated.

Notice also that initially the electron-number was zero, and then the decay created a particle with an electron-number of 1. The total on the right-hand side would be zero if our mystery neutrino had electron-number $= -1$.

Now this *looks* like a mathematical invention. Certainly if the '?' in the electron-number row were -1 then the totals would balance, but this alone does not make it physically correct. Let us for the moment suppose that our mystery neutrino has electron-number -1 and see if that helps to explain any other puzzles.

Another reaction that has been giving us trouble is the β decay of some nuclei:

$$n \to p + e^-. \tag{4.9}$$

This reaction is due to the more basic quark reaction inside the neutron:

$$d \to u + e^-. \tag{4.10}$$

When we discussed this reaction in the previous chapter I suggested that all was not quite as simple as it seemed. If we examine the reaction in terms of lepton number:

$$
\begin{array}{ccccc}
 & d & \to & u & + & e^- \\
L_e & 0 & \neq & 0 & + & 1.
\end{array}
$$

then we see that L_e is not conserved. This dilemma can be solved by suggesting that our mystery neutrino is produced in the reaction as well:

$$
\begin{array}{ccccccc}
d & \rightarrow & u & + & e^- & + & \nu_? \\
L_e \quad 0 & = & 0 & + & 1 & + & -1.
\end{array}
$$

Experimentally it is easy to show that the mystery neutrino is produced in this reaction.

Historically it was in order to solve another puzzle connected with reaction (4.9) that Pauli first suggested the existence of the neutrino type of particle (see section 8.2.4).

Now we have made real progress. The presence of the mystery neutrino in (4.9) was deduced entirely through trying to conserve electron-number. This is concrete evidence that the property exists and is not just an invention of physicists. The next thing to do is see how this neutrino reacts with matter.

We are used to neutrinos reacting with the nuclei of atoms, so we can imagine an experiment in which the mystery neutrinos are allowed to pass through a block of matter in order to see what reactions take place. If this experiment were to be done[2] then something quite remarkable would be discovered. A totally new particle would be produced!

$$\nu_? + p \rightarrow n + e^+. \tag{5.1}$$

The e^+ has a positive electrical charge equal to that of the proton—but it is not a proton as the mass can be measured rather easily and found to be exactly the same as the mass of the electron.

This, as one might imagine, is an important discovery. Historically, the existence of this particle has been known for some time as it was discovered, in rather different circumstances to those described, in 1933[3]. It has been named the *positron*.

Within the context of our discussion, the significance of the positron is that it must have $L_e = -1$ (like the mystery neutrino). This follows from applying the conservation law:

$$
\begin{array}{ccccccc}
\nu_? & + & p & \rightarrow & n & + & e^+ \\
L_e \quad -1 & + & 0 & = & 0 & + & -1.
\end{array}
$$

Notice that the proton and neutron have lepton numbers of 0: they are composed of quarks.

At this point we need to stop and consolidate. This has been a complex section introducing some new ideas and we need to ensure that they have all sunk in before we can go any further.

COFFEE POINT
stop reading
make a cup of coffee
sit and think over the following points

- We have introduced the idea of internal properties to describe the difference between particles that have no other obvious physical difference;
- the internal property that distinguishes the generations of leptons is called lepton number;
- the total lepton number is a conserved quantity in many reactions;
- some reactions do not obviously conserve lepton number;
- we can extend the idea of lepton number by suggesting that the mystery neutrino produced in muon decay has $L_e = -1$;
- this new neutrino also turns up in β decay where it was not expected and helps to solve the generation problem that seemed to occur with this reaction;
- if this new neutrino is passed through matter, then it can react with a proton to produce a positively charged particle with the same mass as the electron—this particle has been named the positron;
- positrons also have electron-number -1.

This is a convincing argument for the physical reality of internal properties.

5.2 Positrons and mystery neutrinos

It seems as if we have introduced a new generation of leptons. As well as the first generation (e^-, ν_e), in which both particles have $L_e = 1$, we have the positron and the mystery neutrino, both of which have $L_e = -1$. However, by definition, a new generation would have to have $L_e = 0$ and a new type of lepton number to identify it. What we have here are particles that are *related* to the first generation, but not standard members of it. They represent a sort of 'inversion' of the first generation.

It is sensible now to stop referring to the 'mystery neutrino': we may as well give it its full name: the *electron antineutrino*, symbolized $\bar{\nu}_e$. The bar over the top of the symbol is to show that it has $L_e = -1$ rather than $+1$ as in the case of the ν_e.

We can now write the full muon decay equation and the full β decay equation:

$$\mu^- \rightarrow e^- + \nu_\mu + \bar{\nu}_e \qquad (5.2)$$

$$n \rightarrow p + e^- + \bar{\nu}_e. \qquad (5.3)$$

We have not yet mentioned the tau decay. Is there also a neutrino with $L_\mu = -1$? A muon antineutrino? Yes, there is!

$$\tau^- \rightarrow \mu^- + \nu_\tau + \bar{\nu}_\mu. \qquad (5.4)$$

The $\bar{\nu}_\mu$ is produced in exactly the same manner as the $\bar{\nu}_e$ in the muon decay. A simple inspection of the muon- and tau-numbers will show that all is well and both are conserved.

Of course, we are now led to the next question: what happens when $\bar{\nu}_\mu$ passes through matter?

$$\bar{\nu}_\mu + p \rightarrow n + \mu^+. \qquad (5.5)$$

As one might expect, there is a new particle produced which is exactly the same mass as the muon, but with a positive electrical charge. It is called the *antimuon*.

The pattern, which is shown in table 5.2, is completed by the *antitau* and the *tau antineutrino*.

In this table 'lepton number $= -1$' should be taken to mean that either the electron-number or muon-number or tau-number $= -1$, depending on the generation.

By this time the reader may be feeling somewhat aggrieved. In chapter 1 I listed six material particles from which all matter in the universe is

Table 5.2. The extended lepton families.

	1st generation	2nd generation	3rd generation
lepton number $= +1$	$\begin{bmatrix} e^- \\ \nu_e \end{bmatrix}$	$\begin{bmatrix} \mu^- \\ \nu_\mu \end{bmatrix}$	$\begin{bmatrix} \tau^- \\ \nu_\tau \end{bmatrix}$
lepton number $= -1$	$\begin{bmatrix} e^+ \\ \bar{\nu}_e \end{bmatrix}$	$\begin{bmatrix} \mu^+ \\ \bar{\nu}_\mu \end{bmatrix}$	$\begin{bmatrix} \tau^+ \\ \bar{\nu}_\tau \end{bmatrix}$

made, and now, four chapters further on, we are discussing six more objects that were not present in that list.

Technically speaking the list in chapter 1 was perfectly complete—it includes all the fundamental *matter* particles. These new objects, with lepton numbers $= -1$, are known as *antimatter* particles.

If you thought that antimatter was the creation of science fiction and only existed in the engine room of the Starship Enterprise, then I am afraid that you are wrong. Antimatter exists. It can be created in the lab and particle physicists[4] frequently use it in their experiments. Some of the stranger properties that writers tend to give antimatter (like it producing antigravity) are not true, however (antiparticles fall in a gravitational field, they do not rise!).

Antimatter particles have the same mass as the equivalent particle, opposite charge, *if the particle has charge*, and, at least in the case of the leptons, opposite lepton number.

Antimatter is remarkably rare in the universe. Nobody is totally sure why the universe does not consist of equal amounts of matter and antimatter. If there is a large amount of antimatter in the universe, then it is hidden somewhere[5].

In summary, we have six leptons split into three generations and six antileptons also split into three generations. The antileptons reflect the properties of the leptons, most obviously by a reversal of charge and lepton number.

5.3 Antiquarks

We arrived at the idea of antileptons through the introduction of the three lepton numbers that distinguish between the generations. Perhaps we can extend the idea to quarks by introducing three quark numbers to distinguish the quark generations?

We could define a 'down-number' that the d and u quarks have, a 'strange-number' that the s and c quarks have and a 'bottom-number'(!) possessed by b and t. Unfortunately, this does not work.

If you remember, the weak force does not totally separate the quark generations as it does the lepton generations. This would imply that the 'up-number', etc, would not be conserved quantities in the weak interaction, hence they would not pass one of the tests that we applied to lepton number to be sure that it was a real property. One could introduce these numbers, they just would not tell us anything useful about *all* weak reactions.

The situation is not totally lost. There is a property that distinguishes the quarks—that they are not leptons! The weak force does *absolutely* distinguish quarks from leptons; it cannot turn a quark into a lepton.

Another way of seeing this is to remember that all the quarks have all three lepton numbers $= 0$. This is equivalent to not having a lepton number 'switch'. Although quarks do not carry lepton number, they do carry a different form of 'internal switch'—one that makes them quarks rather than leptons. This internal property is referred to as *baryon number* (B). All quarks have baryon number $= 1/3$, all leptons have baryon number $= 0$. The weak force, as well as the other forces, conserves baryon number:

CONSERVATION OF BARYON NUMBER
In all reactions the total baryon number of the particles
before the reaction must be the same as the total baryon number
after the reaction.

If the leptons were not strictly divided into generations by the weak force, then there would only be a single lepton number rather than three, and the situation for the quarks and leptons would be identical.

Two things seem odd straight away. Why is the property called baryon number not quark number, and why do quarks have baryon number = 1/3 rather than 1?

Baryon number was invented before quarks were discovered. Its name was coined independently from the name 'quark'. We shall see where the name comes from in chapter 6. Quarks carry $B = 1/3$ because baryon number was originally defined with the proton in mind. The proton is assigned baryon number = +1. As protons contain three quarks, each quark has baryon number = 1/3: the total baryon number of the proton is the sum of the baryon numbers of the particles from which it is composed.

This is a subtle point that needs some further discussion. Baryon number, and indeed the lepton numbers, are supposed to represent internal properties of particles. Such properties do not have a size in the sense that electrical charge or mass have a size. This being the case, *it does not matter what number we use to represent them.* There would be no problem in saying that the electron has $L_e = -42.78$, as long as we also said that the ν_e had $L_e = -42.78$ and that the e^+ had $L_e = +42.78$. It would be a rather silly number to deal with, but the size of the number is irrelevant. When we represent binary numbers in electronics we use a similar convention. For example, binary 1 is taken as +5 volts and binary 0 as 0 volts. In principle any pattern of voltage would have done the job.

The pattern we have chosen for baryon number is based on calling the proton the $B = +1$ particle. The only possibilities for the baryon numbers of *fundamental* particles are $B = +1/3, 0, -1/3$. $B = +1/3$ are the quarks, $B = 0$ are the leptons and $B = -1/3$ are the *antiquarks*.

Just as for the leptons, each quark has an antiquark partner. Each antiquark has the opposite charge but the same mass as its partner matter quark.

Putting all this into a table like that for the leptons we obtain table 5.3.

The antiquarks are symbolized, as in table 5.3, by the equivalent quark symbol with a '-' or bar over the top. An antiup quark is written $\bar{u}$ and pronounced 'u bar'.

Table 5.3. The extended quark families.

	1st generation	2nd generation	3rd generation
baryon number $= +\frac{1}{3}$	$\begin{bmatrix} u \\ d \end{bmatrix}$	$\begin{bmatrix} c \\ s \end{bmatrix}$	$\begin{bmatrix} t \\ b \end{bmatrix}$
baryon number $= -\frac{1}{3}$	$\begin{bmatrix} \bar{u} \\ \bar{d} \end{bmatrix}$	$\begin{bmatrix} \bar{c} \\ \bar{s} \end{bmatrix}$	$\begin{bmatrix} \bar{t} \\ \bar{b} \end{bmatrix}$

The strong interaction acts between antiquarks just as it does between quarks, implying that antiquarks combine to form antiparticles. The most obvious examples of this are the *antiproton*, a $\bar{u}\bar{u}\bar{d}$ combination, and the *antineutron* $\bar{n}$, $\bar{u}\bar{d}\bar{d}$.

The antiproton will have the same mass as the proton, but a negative electrical charge. The antineutron, however, will still have a zero charge, but is different from the neutron (despite having the same charge and mass) because it is composed of antiquarks rather than quarks.

We shall return to the subject of combinations of antiquarks when we have discussed in more detail the combinations of quarks—something that we will do in chapter 6.

5.4   The general nature of antimatter

So far we have introduced antileptons as particles with negative lepton numbers and antiquarks as having negative baryon numbers. The justification for doing this is that it helps us make rules about lepton and baryon number conservation. Of course, this would just be playing with numbers if it were not for the fact that these antiparticles exist and can be created in the lab. However, there is far more to antimatter than this simple introduction would suggest.

The first hint of the existence of antimatter came, in 1928, as a consequence of the work of the English theoretical physicist P A M Dirac. Dirac had been doing research into the equations that govern the motion of electrons in electric and magnetic fields. At slow speeds the physics was well understood, but the discovery of relativity in 1905 suggested that the theory was bound to go wrong if the electrons were moving at speeds close to the speed of light.

Dirac was the first person to guess the mathematical equation that correctly predicted the motion in these circumstances[6]. The Dirac equation, as it is now known, has become a vital cornerstone of all theoretical research in particle physics. Unfortunately, the mathematics that we would need to learn to be able to deal with this equation would take far too long to explain in a book like this.

When Dirac came to solve his equation he discovered that it produced two solutions (in the same way that an equation like $ax^2+bx+c = 0$ will produce two solutions). One solution described the electron perfectly, the other a particle with the same mass but opposite electrical charge. This was a considerable worry to Dirac as no such particle was known, or suspected, at the time.

No matter what mathematical tricks Dirac tried to pull, he could not get rid of this other solution. At first he thought that the second solution must represent the proton, but the huge difference in mass made this untenable. The problem was resolved when Anderson discovered the positron in 1933. This particle was the second solution to Dirac's equation.

Dirac unknowingly predicted the existence of antimatter. Since that time many famous physicists have worked on the general theory of antimatter[7].

We now know that the existence of antimatter follows from some very basic assumptions about the nature of space and time. The existence of both matter and antimatter is necessary, given the universe in which we exist. This is a very deep and beautiful part of particle physics.

Any theory that is consistent with relativity must contain both particles and antiparticles.

Unfortunately I can think of no adequate way of explaining this at the level that this book intends. I must content myself with saying that the very basis on which the universe is set up requires that every fundamental particle must have an antiparticle partner[8].

5.5 Annihilation reactions

Science fiction writers have used antimatter for many years, so it is hardly surprising that some people's knowledge is influenced by such stories. Unfortunately, the writer does not always stick to the truth when it suits the story better to bend the facts. However, one fact is accurately portrayed: matter and antimatter will destroy each other if they come into contact.

No-one has ever mixed large amounts of antimatter and matter together to produce an explosion of atomic bomb proportions. This is because antimatter is difficult to make in anything other than minute amounts[9].

However, particle physicists regularly collide small amounts of matter and antimatter together because of the energy and new particles that can be produced in such reactions. These are called *annihilation reactions*. A typical example of an annihilation reaction is:

$$e^+ + e^- \rightarrow ?$$

We shall come to understand how important this reaction has been for the development of particle physics.

Let us, for the moment, not worry about the practical details of how to arrange for such a reaction to take place and how to detect the particles that are produced. Instead we need to examine the fundamental features that emerge. Imagine that we have two beams, one of e^+ and the other of e^-, that we can allow to come together at will with any energy we choose. Examining the reactions that take place when the beams collide reveals several interesting features.

- At all energies the e^+ and the e^- can simply bounce off each other without anything interesting happening.
- If the e^+ and the e^- are moving very slowly, then they can go into orbit round each other producing an atom-like object called positronium. Positronium is very unstable: the e^+ and the e^- will invariably destroy each other.
- If we increase the reaction energy, i.e. the kinetic energy of the e^+ and the e^- entering the reaction, then something interesting happens. Once we have exceeded a precise threshold energy, the following reaction starts to take place:

$$e^+ + e^- \rightarrow \mu^+ + \mu^-. \tag{5.6}$$

Inspection of the lepton numbers soon indicates that all is well; the reaction does not violate any conservation law. However, it is quite a surprising thing to see happening.

- If the reaction energy is turned up some more, then both reactions continue to happen until another threshold is passed, at which point another reaction[10]:

$$e^+ + e^- \rightarrow \tau^+ + \tau^- \tag{5.7}$$

starts to take place.

- In between these thresholds some other boundaries are passed as well. As the energy is increased particles are produced indicative of the presence of quarks. These 'jet' reactions, as they are called, are not as 'clean' as the lepton reactions. The strong interaction makes them complicated. For the moment we shall not study them. They will form an important section of chapter 9, as they represent direct evidence for the existence of quarks.

When the e^+ and the e^- react together the initial combination of particles is such that the electrical charge, lepton number and baryon number all total to zero. This allows the reaction to produce anything, provided the combination adds to zero again. Any generation of lepton, or any generation of quark, can be produced provided the appropriate antiparticle is produced as well. Hence, annihilation reactions are an easy way of searching for new particles.

The materialization of particles and antiparticles is a consequence of the relationship between mass and energy contained in the theory of relativity. There must be enough energy in the reaction to produce the pair of particles. For example, the total energy of the e^+ and e^- needed to produce a tau pair is approximately 16.96 times that required to produce a muon pair as the mass of the tau is 16.96 times the mass of a muon.

This is such an important point that it needs summarizing:

> Annihilation reactions between matter and antimatter can lead to the production of new matter and antimatter provided they are produced in equal amounts and provided there is enough energy in the reaction.

This annihilation of matter and antimatter, leading to the creation of new matter and antimatter, seems to imply an essential symmetry between them. Every time matter and antimatter annihilate each other some new combination of matter and antimatter is produced, so it is difficult to see how there could be different amounts of matter to antimatter in the universe. If there are equal quantities of matter and antimatter, then where is all the antimatter? This returns us to one of the fundamental problems in cosmology—the creation of the universe would violate our conservation laws, unless equal amounts of matter and antimatter are produced!

5.6   Summary of chapter 5

- There are three types of lepton number which are conserved in all reactions;
- each type of lepton number is characteristic of one of the generations;
- the lepton number represents a new, internal, property of leptons;
- particles with negative lepton numbers exist and are called antileptons;
- for each lepton there is an equivalent antilepton;
- a charged lepton and its antilepton partner have the same mass but opposite electrical charge and lepton number;
- neutral leptons and antileptons differ in their lepton numbers;
- quarks have a property called baryon number that is conserved in all reactions;
- baryon number is not generation specific;
- every quark has an antiquark partner with opposite charge and baryon number;
- matter/antimatter annihilation reactions lead to new forms of matter being produced;
- these annihilation reactions point to a relationship between matter and energy.

Notes

[1] There are various terms used for internal properties: they are sometimes referred to as intrinsic properties, or quantum numbers.

[2] Indeed, this experiment can be done, it is just that the neutrinos are produced in a different way.

[3] Its discoverer, Anderson, won the Nobel Prize for finding the positron in a cosmic ray experiment.

[4] Even the ones without pointed ears!

[5] It is difficult to see how large amounts of antimatter could be hidden. Presumably, it would have to be in some separate region of the universe. However, at the boundary between the matter and antimatter regions reactions would take place that would produce enormous quantities of gamma rays. These would be easily observed by astronomers, but have not been seen.

[6] Dirac was awarded the Nobel Prize in 1933.

[7] This is why the positron is not called the antielectron. When it was discovered, physicists had no idea of the general nature of antimatter.

[8] This is not to say that both must exist at the same time. Theory says that each particle *type* must also have an antiparticle *type*, not that every particle in the universe at this moment must have an antiparticle partner somewhere.

[9] In September 1995 a team lead by Professor W Oelert working at CERN managed to produce atoms of antihydrogen—antiprotons with antielectrons in orbit. The experiment used the low energy antiproton ring (LEAR) at CERN. During its three week run, the experiment detected nine anti-atoms. See Interlude 2 for more information.

[10] This is how the tau was discovered!

Chapter 6

Hadrons

In this chapter we shall study quarks and their properties in more detail. This is a more complicated task than that for the leptons. All quarks and antiquarks are bound into composite particles by the strong interaction. In order to study the quarks, we must study these particles. This chapter is the first of three closely related chapters in which we shall study the properties, reactions and decays of hadrons.

6.1 The properties of the quarks

It is not possible to present a table of the physical properties of the quarks with the same degree of confidence as we did for the leptons in chapter 4. The electrical charges of quarks can be stated with some certainty, but the masses of the quarks are a source of controversy.

6.1.1 Quark masses

Many textbooks quote values of quark masses that are similar to those in table 6.1.

All such tables should be taken with a pinch of salt. Other textbooks quote quark masses that are very different from the ones I have chosen to use. The values of masses that are used depend on the calculations that are being carried out. The values of mass that I have presented in table 6.1 are those that are most appropriate for the use to which we will put them.

Table 6.1. The masses of the quarks (in GeV/c^2).

Charge	1st generation	2nd generation	3rd generation
+2/3	up	charm	top
	0.33	1.58	175[1]
−1/3	down	strange	bottom
	∼0.33	0.47	4.58

The key issue here is the nature of the strong force. The strong force binds quarks together into composite particles. We believe that it is theoretically impossible to isolate an individual quark (or antiquark). This makes it impossible to directly measure the mass of an individual quark. Quark masses have to be deduced from the masses of the composite particles that they form. However, this is not as simple a task as it sounds.

The strong force holds the quarks together tightly, so very large energies are involved inside the particles. As we know from chapter 2, energy implies mass so some of the mass that we measure for a composite particle is due to the masses of the quarks, and some is due to the energy of the forces between the quarks. This makes deducing the masses of the individual quarks very difficult. Imagine that you were trying to estimate the mass of an ordinary house brick, and all the information that you had was the masses of various houses. Even if you knew exactly how many bricks there were in each house, the bricks are of different types and you need to take into account all the other materials that go to make up the house as well. Such is the task of the physicist trying to deduce quark masses.

At the moment it is impossible to calculate how much of the mass of a proton, say, is due to the masses of the quarks inside and how much is due to the interaction energies between the quarks. On the other hand, when these composite particles interact with one another the result is often due to the interactions of the individual quarks within the particles, hence the masses of the quarks have some influence on the reaction. This gives us another way of calculating the quark masses.

Unfortunately, these two techniques result in very different values for the quark masses. At the moment we do not have a complete mathematical understanding of the theory, so the puzzle cannot be resolved.

The values that are quoted in table 6.1 are values deduced from the masses of composite particles. Given the degree of uncertainly associated with table 6.1, what can we reasonably deduce from it? The most important point to gain from this table is the *relative values* of the quark masses. The antiquark masses are assumed to be the same as the quark masses: the charges are, of course, reversed.

The similarity in the masses of the u and d quarks is well established experimentally. After all the proton (uud) and the neutron (udd) are very nearly the same mass. Other particles composed of u and d combinations also have small mass differences, which emphasizes how close the u and d are. The d quark is slightly more massive than the u—if this were not the case neutrons (udd) would not be unstable particles that decay into protons (uud).

Charm, strange and bottom are now quite familiar quarks although our experiments have only recently been of high enough energy to produce enough b quarks for them to be systematically studied.

The chronology of discovery is interesting. The c quark was a surprise discovery in 1974 and the b came soon after in 1977. However, despite great effort by experimenters we had to wait until 1994/95 for the top quark to be discovered. This is a reflection of the much greater mass of this quark. Experiments had to reach much higher energies to produce particles that contained t quarks.

6.1.2   Internal properties of quarks

We came across one internal property of quarks in chapter 5—baryon number. All quarks have baryon number $= 1/3$ and all antiquarks have baryon number $= -1/3$. Strictly speaking, baryon number is the only internal property of quarks. They do not have quark numbers like the lepton numbers, because the weak force does not respect the quark generations.

Physicists have found it useful to label quarks with 'properties' that do not pass all our tests for internal properties (they are not conserved by the

Table 6.2. The quark flavour numbers.

Quark	U	D	C	S	T	B
up	1	0	0	0	0	0
down	0	−1	0	0	0	0
charm	0	0	1	0	0	0
strange	0	0	0	−1	0	0
top	0	0	0	0	1	0
bottom	0	0	0	0	0	−1

weak force), but work quite well for the other forces. These are useful as they help us to keep track of what might happen when quarks interact. These properties are not referred to as quark numbers but as 'flavour numbers'.

Flavour is another whimsical term that is commonly used in the particle physics community. Particle physicists like to refer to the quarks as having different flavours—up, down, strange, etc. There are six flavours of quarks, and six flavours of leptons. Table 6.2 lists the assignment of flavour numbers. Note that the flavour numbers belong to *each quark*, not to *each generation* as in the case of the leptons. Electron number is a property of the e^- *and* the ν_e. There is no common flavour number to the u and d quarks. Each quark has its own distinctive flavour number.

The antiquarks also have flavour numbers, the values being the negative of the corresponding particle flavour number, i.e. $\bar{c}$ has $C = -1$. The symbols S and C refer to 'strangeness' and 'charm' (not strangeness-number: although this name was used, it rapidly went out of fashion). T and B are 'topness' and 'bottomness'. There was a strong lobby for 'truth' and 'beauty' by the romantic element of the particle physics community, but the names did not gain much support.

U and D should refer to 'upness' and 'downness', but as these properties are hardly ever used they have not gained familiar names. It has never proved useful to use U and D numbers as the up and down quarks are so similar in mass. U and D have been included for completeness only.

These flavour numbers are very useful for keeping track of what happens to the various quarks when particles react together. We shall use them extensively in this way in the next chapter.

It may seem odd that some of the flavour numbers are -1. For example the strange quark has $S = -1$ (which means that the antistrange quark has $S = +1$). This is a consequence of the properties being defined *before* the discovery of quarks, as baryon number was. Unfortunately, the particle that was first assigned strangeness $+1$ contained the $\bar{s}$ quark, not the s. Hence we are stuck with the s quark having $S = -1$.

Before we can use the flavour numbers to study the reactions of composite particles, we must understand the basic features of the strong force that binds the quarks into these particles.

6.2   A review of the strong force

The strong force only exists between quarks. The leptons do not feel the strong force.

The strength of the force varies with distance, much more so than gravity or the electromagnetic force. The strong force effectively disappears once quarks are separated by more than 10^{-15} m (about the diameter of a proton). For distances smaller than this, the force is very strong indeed. The exact details of how the force's strength changes with distance are irrelevant at this stage and so we will postpone them until chapter 7.

The strength of the force does not depend on the type of quark that is involved, nor on the charge of the quark. Physicists say that the strong force is 'flavour independent'.

For the moment, it is best to think of the strong force as being purely attractive and so strong that it binds the quarks together in such a way as they can never escape. However, the properties of the strong force are such that the quarks don't *all* stick together in one large mass (otherwise the universe would be a huge lump of quarks). The strong force ensures that quarks and antiquarks can only stick together in groups of three (qqq or $\bar{q}\bar{q}\bar{q}$), or as a quark and an antiquark pair ($q\bar{q}$).

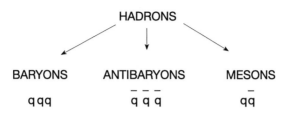

Figure 6.1. The hadron family tree.

6.3 Baryons and mesons

The family name for composite particles made from quarks is *hadrons*. Before quarks were discovered, particle physicists thought that leptons and hadrons were the only constituents of the universe.

The hadron family tree has three branches—those that consist of three quarks bound together, those that are three antiquarks and those that consist of a quark and an antiquark.

A particle that is composed of three quarks is a member of the sub-group called *baryons*. If it is composed of three antiquarks, then it is an *antibaryon*. The quark–antiquark pairs are called *mesons*. The proton and neutron (uud and udd) are baryons, the π^+—pronounced 'pi plus'—(ud̄) is a meson. Figure 6.1 shows the various categories of the hadrons.

Most of what we have to say about the baryons is also true about the antibaryons, except for obvious things like an antibaryon having the opposite electrical charge to the similar baryon, so we shall not refer to them explicitly unless we need to point out a different feature.

All baryons, except the proton, are unstable and will decay. The particles produced must include another baryon. Consequently, decay chains are formed as one baryon decays into another that is also unstable, etc. Such decay sequences must end at the proton, which is the lightest baryon and therefore stable.

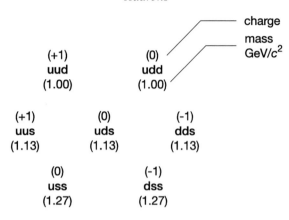

Figure 6.2. Quark combinations (predicted particle masses in GeV/c^2).

6.4　Baryon families

If we restrict ourselves for the moment to just the lightest three quarks, u, d and s, then the number of possible ways in which they can be combined is a small enough number to be manageable. When the combinations are listed the order of the quarks is not important—uds would be the same as sud or dus. (By the way, we do not try to pronounce these as words, 'sud—a particle often met in washing machines', the individual letters are pronounced.)

Figure 6.2 shows a selection of the possible combinations with their predicted masses in GeV/c^2. For the moment I have missed out uuu, ddd and sss. It is quite easy to use the properties of the quarks to deduce what baryons made from these combinations should be like. For example, the uus combination should have an electrical charge $= (+2/3) + (+2/3) + (-1/3) = 1$. Its mass should be $= (0.33 + 0.33 + 0.47)$ GeV/$c^2 = 1.13$ GeV/c^2. These properties have already been listed in figure 6.2, the charge above the combination and the mass below.

As one might expect, there is a gradual increase in the mass as the number of strange quarks increases. When we examine the actual particles that are formed we might also expect a small increase in mass along horizontal rows of the figure as the d quark is slightly more massive than the u.

		strangeness	name
p (0.938)	n (0.940)	S=0	nucleon
Σ^+ (1.189) Σ^0 (1.192) Σ^- (1.197)		S=-1	sigma
Ξ^0 (1.314) Ξ^- (1.321)		S=-2	cascade

Figure 6.3. A baryon weight diagram (particle masses in GeV/c^2).

The horizontal rows form *baryon families*. A family is defined as a set of baryons of similar mass with the same internal properties. The top row forms a family as the two particles have the same mass and the same strangeness ($S = 0$). The second row is a family of similar mass all with $S = -1$ and the bottom row the $S = -2$ family. This pattern of particle properties was noticed before the quark structure of baryons was known about. Indeed it was one of the important pieces of evidence that led Murray Gell-Mann and George Zweig to the discovery of quarks.

Figure 6.3 is a chart of various particles that have been discovered in reactions. Such a chart is sometimes called a *weight diagram* and they can be plotted in various ways. In this case I have grouped the baryons into family sets with mass increasing down the diagram. The diagram clearly follows the pattern of the quark combinations very well.

The mass predictions have not turned out to be exactly right, but this is to be expected given the uncertainly in the masses of the quarks and the nature of the forces between them. However, the pattern and its correspondence to that of the quark combinations are compelling.

There is one known baryon that is missing from this weight diagram. The lambda baryon (Λ) is known to be strangeness -1 (this can be deduced from the way it is produced) and has a mass of 1.115 GeV/c^2. This makes it a strong candidate to be a uds combination—the mass is about right and the strangeness value points to the presence of only one s quark. However, the uds combination would seem to be already taken by the Σ^0. In fact there is no reason why both particles should not have the same quark content.

One of the problems associated with calculating the masses of baryons is that the quarks inside have a variety of energy levels that they can occupy. This is a similar situation to that found in atoms in which the electrons in orbit round the nucleus have a variety of energy levels (see section 3.6.5). In an atom there are restrictions that must apply—no more than two electrons can be in the same energy level at any one time for example. Similar restrictions apply to quarks in baryons (and mesons) except the situation is more complicated as the energy levels are mostly determined by the strong forces between the quarks. In the case of the baryons and mesons the energy levels of the quarks have a significant influence on the mass of the particle—electronic energy levels hardly contribute to the mass of an atom.

Given a quark combination like uds there is greater freedom in how the quarks can be organized within the energy levels. The Σ^0 and the Λ masses reflect this—they have the same quark content but different masses as the quarks are in different energy levels. The Σ^+ and Σ^- have two identical quarks (Σ^+ uus, Σ^- dds) which restricts their energy levels more; they do not have a duplicate lower mass version like the Λ.

The complete weight diagram including the Λ is known as the *baryon octet*. This consists of quark combinations in low mass states. There is a similar set of particles in which the quarks occupy higher energy level states. Their properties are essentially the same as their lower mass brothers, except that their greater intrinsic energies make them more prone to decay.

The uuu, ddd and sss combinations can *only* exist in these higher mass states. Having three identical quarks prevents them from occupying energy levels in such a way as to produce low mass versions. This is why I omitted these combinations in figure 6.2. In figure 6.4 the set of higher mass particles is grouped into a weight diagram. Notice that the delta family contains two particles which have exactly the same quark combinations as protons and neutrons—the Δ^+ is a uud, like the proton and the Δ^0 is a udd, like the neutron. Essentially these are higher mass versions of the proton and neutron. However, the delta family goes on

				strangeness	name
Δ^-	Δ^0	Δ^+	Δ^{++}	S=0	delta
ddd	udd	uud	uuu		
(1.23)	(1.23)	(1.23)	(1.23)		
	Σ^{*+}	Σ^{*0}	Σ^{*-}	S=-1	sigma*
	dds	uds	uus		
	(1.383)	(1.384)	(1.387)		
		Ξ^{*0}	Ξ^{*-}	S=-2	cascade*
		uss	dss		
		(1.532)	(1.535)		
		Ω^-		S=-3	omega
		sss			
		(1.67)			

Figure 6.4. The higher mass baryons.

to include the Δ^{++} which is uuu and the Δ^- which is ddd. Neither of these combinations can be found in the lower mass table.

This diagram is known as the *baryon decuplet*. Organizing the baryons and tabulating their properties in this way was named 'the eightfold way' by Murray Gell-Mann (who else?) after the Zen Buddhist path to enlightenment. Using his analysis, which eventually led him to suspect the existence of quarks, Gell-Mann *predicted* the existence of the Ω^- (sss) particle, which was discovered in 1964 with exactly the properties and mass that Gell-Mann suggested[2]. The sss does not occur in the octet diagram either so the Ω^- has no lower mass brother.

6.4.1 Higher mass baryons

Having spent some time discussing different baryons, it is rather sobering to realize that we have only employed half of the quarks that are at nature's disposal in the creation of particles. However, we have covered all the basic physics.

Hadrons

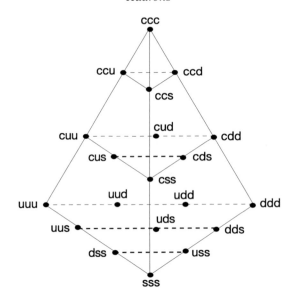

Figure 6.5. A charming extension to the baryon decuplet.

Simply adding the charm quark to our game adds enormously to the complexity. For example, the baryon decuplet becomes a three-dimensional pyramid. The base of the pyramid is formed from the decuplet, and the point at the top of the pyramid is the ccc combination. The full pyramid of combinations is illustrated in figure 6.5.

Of course, a similar extension must be made to the baryon octet as well (without the ccc combination).

At this point we start to run into the limits of our experimental knowledge. Baryons that contain the heavy quarks require a great deal of energy to produce—more than can be provided in many experiments. The Λ_c^+, which is udc and the Σ_c^{++}, which is uuc, were discovered in 1975 and the Λ_b (udb) in 1981. The study of these heavy flavour baryons is still being advanced.

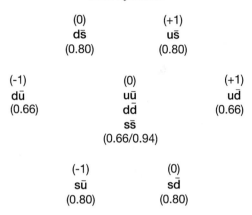

Figure 6.6. Meson quark combinations.

6.5 Meson families

A meson is a hadron that consists of a quark–antiquark pair. The quark and the antiquark do not have to be of the same flavour. Any combination is possible, although some are less stable than others.

The u, d and s quarks (with their antiquarks) are the least massive quarks, and hence the most common in nature, so we might reasonably expect that mesons formed from them would also be quite common.

In figure 6.6 strangeness decreases as one moves down the diagram, but, in contrast to the baryon diagrams, mass does not increase. As expected there is a set of mesons that corresponds to these combinations (figure 6.7).

This time, as there are nine of them, the set is called the *meson nonet*.

The central row of three particles, the π^-, π^0 and π^+ make up the pion family, the lightest mass and most common of the mesons. Occupying the same slot as the π^0 are two other particles, the η and the η'.

The π^0 is the uū *and* the dd̄ combination. This does *not* mean that there are two types of π^0, one containing uū and the other containing dd̄ . As explained in chapter 8, the uū combination can change into the dd̄ combination, and back again, inside the particle.

			strangeness	family name
K⁰		K⁺	S=+1	kaons
(0.498)		(0.494)		
π⁻	π⁰/η/η′	π⁺	S=0	pions/eta
(0.140)	(0.135/0.547/0.958)	(0.140)		
K⁻		K̄⁰	S=-1	kaons
(0.494)		(0.498)		

Figure 6.7. The meson nonet.

The η and the η' are different combinations of u$\bar{\text{u}}$, d$\bar{\text{d}}$ and s$\bar{\text{s}}$ states.

The kaon family is split on this diagram as the four particles have different strangeness. The K⁺ and the K⁻ are antiparticles of each other. One of the advantages of this diagram is that both types of particle are displayed on one picture.

Note that unlike the π^0, the K⁰ and K̄⁰ are different particles with slightly different properties. This seems quite strange (no pun intended). Both mesons have a zero electrical charge and the same mass, so what is the difference between them? The differences lie in the reactions that the two neutral kaons can take part in and the ways in which they decay. The situation is not so very different from that of the neutron. Its antiparticle has the same mass and the same zero electrical charge, but is a distinct object that reacts in different ways.

The situation is even more odd in the case of the neutrinos; their antiparticles are only different from the particle by having a different value of the lepton number and that is not something that can be measured directly—see chapter 5.

The masses of the particles have turned out to be rather less than one might expect given the quark masses from the previous section. This is typical of the problems associated with giving quarks mass. Values that are chosen do not seem consistent with all particles. This problem arises from the different ways in which the strong force binds the quarks together in a baryon (qqq) and a meson (q$\bar{\text{q}}$).

Other, less common, mesons containing heavier quarks have been created and studied in particle physics experiments. They are difficult to create, because of their large masses and their study, although much better developed than the study of heavy baryons, is still continuing.

6.6 Internal properties of particles

Particle physicists have chosen to label various hadrons by the internal properties of the quarks that they contain. To give an immediate example, any hadron that contains a strange quark is called a strange particle and given a value of the internal property S.

Strictly, S is a property that can only belong to a quark, but this convention was developed *before* quarks were discovered so, historically, physicists are more used to giving hadrons S values than they are quarks[3].

The K$^+$ was the particle chosen to define the values of S for all the others. By convention the K$^+$ is given $S = +1$. Notice that the quark combination of the K$^+$ is u$\bar{\text{s}}$—this is why it is the $\bar{\text{s}}$ that has $S = +1$, not the s quark, as might seem more sensible.

Despite its apparent eccentricity (in hindsight) the labelling of particles as if they had internal properties is very useful. The S value of the particle tells us how many strange quarks it contains (OK, it counts them in -1's, but nothing is perfect!).

For example, the Ξ^- has $S = -2$, so it must contain two strange quarks. As S values are quoted in particle data tables, we can work out what the quark content of a particle is by looking up a couple of numbers in a table.

To give a full example, let us work out the quark content of the Σ^+. From a data table (e.g. the back of this book):

(1) the Σ^+ is a baryon;
(2) the Σ^+ is a strange particle, with $S = -1$;

(3) the Σ^+ has a charge of $+1$.

The first point tells us that the Σ^+ contains three quarks (as that is the definition of a baryon). The second point implies that it contains a strange quark. In addition, the fact that it does not have any other internal properties (i.e. C, T or B) implies that the other quarks must be u or d. Noting from the table that the s quark has charge $-1/3$, we can work out that the other quarks in the Σ^+ must have a total charge of $+4/3$ in order to make the particle $+1$. The only way to make this from two quarks is uu. Hence the Σ^+ is an suu combination.

Other internal properties exist as well. The ones that are listed are C, B and T which cover particles that contain the charm, bottom and top quarks. For example the D^0 meson is a cū combination and so has $C = +1$. These internal properties count the respective quarks just as well as S does.

6.7   Summary of chapter 6

- It is difficult to give the quarks unambiguous masses;
- quarks have flavour numbers which are not really internal properties but which are quite useful 'book keeping' devices;
- particles composed of quarks are called hadrons;
- a meson is a q$\bar{\text{q}}$ combination;
- a baryon (antibaryon) is a qqq ($\bar{\text{q}}\bar{\text{q}}\bar{\text{q}}$) combination;
- there is a sequence of baryon families based on the proton and neutron;
- this sequence forms the baryon octet with increasing numbers of strange quarks;
- there is a sequence of baryon families (the decuplet) based on the Δ particles which have greater masses than the octet families;
- there are more particles in the decuplet than the octet as the symmetrical combinations (uuu, ddd, sss) are allowed in the decuplet but not in the octet;
- particles of different masses can have the same quark content as the quarks can be in different energy levels (which gives rise to the different mass);
- the symmetrical combinations can only occur if some of the quarks are in higher energy levels;

- there is a meson nonet which has a similar structure to the baryon octet;
- hadrons are tabulated in particle tables as having values of internal properties such as S, C, B and T;
- these are actually the flavour numbers of some of the quarks they contain;
- these internal properties of particles are useful clues to the quarks that they contain.

Notes

[1] Notice that this fundamental particle is more massive than many decent sized molecules!

[2] The Ω^- prediction was made at a conference in CERN in 1962 by Gell-Mann and, independently, by Yuval Ne'eman. Unfortunately for Ne'eman, Gell-Mann was called to the blackboard before him.

[3] In a way this is a very good use of the term internal properties as the properties belong to objects inside the particle. Really and truly, they should be called flavour numbers as they are no more internal properties than the flavour numbers of quarks are. The only genuine internal properties we have come across are the lepton numbers and baryon number.

Chapter 7

Hadron reactions

In this chapter we shall use what we have learnt about their quark composition to discuss how hadrons react with each other. In the process we shall learn about various rules that help us to decide which reactions are physically possible. These rules, called conservation laws, are intimately related to the way in which quarks and leptons are acted upon by the fundamental forces.

7.1 Basic ideas

Particle reactions take place when two or more particles come close enough to each other for fundamental forces to cause an interaction between them. As a result of this interaction two things may happen:

(1) the particle's trajectory can change: they can be attracted, or repelled;
(2) the particles can change into different particles: it may be that a greater number of particles emerge from the reaction than entered it.

Of course, these possibilities are not mutually exclusive: if more particles are created, then the trajectories must have altered.

The fundamental forces at work and the energy of the reacting particles control the exact outcome of the reaction. As the energy increases a wider range of possibilities becomes open to the forces. Observing how the various possible reactions behave at different energies is an important technique in studying the fundamental forces.

The study of hadron reactions was the focus of particle physics in the 1950s and 1960s. As technology improved, experimenters collided particles at higher and higher energies. Physicists soon realized that the intrinsic and kinetic energies of the incoming particles were being mixed and redistributed by the fundamental forces. From this mixture new particles were materializing.

7.2 Basic processes

When two hadrons react with each other to form new particles it is invariably due to the strong force as:

- the gravitational force is far to weak to have any effect on fundamental particles;
- the weak force is much shorter range than the strong force, so a strong force reaction will tend to take place before the hadrons get close enough together for the weak force to come into play;
- the electromagnetic force is much longer range than either the weak or strong forces so hadrons coming together may well attract or repel each other via the electromagnetic force but at a distance; if they have enough energy to approach each other closely they will react via the strong force.

Early experiments collided protons with protons. This is quite easy to arrange experimentally. One technique is to use a tank of liquid hydrogen (called a bubble chamber) and to fire a beam of protons into the tank. Liquid hydrogen consists of protons and orbiting electrons. The electrons very rarely react with the protons passing through the liquid, so the tank is basically a collection of protons that will react with those in the beam, if they get close enough.

The most obvious reaction between two protons is:

$$p + p \rightarrow p + p \qquad (7.1)$$

in which they simply bounce off each other (elastically scatter). This reaction can take place at any energy and is a consequence either of the electrostatic repulsion between the two positively charged protons, or of a strong force reaction which has not broken up the particles. As the energy in the reaction is increased (this corresponds to increasing the speed of the incoming protons) other strong force reactions start to take place such as:

$$p + p \rightarrow p + p + \pi^0. \tag{7.2}$$

Considering reaction (7.2) in more detail, one thing becomes apparent immediately: the total number of quarks and antiquarks in the reaction has changed:

p	+	p	→	p	+	p	+	π^0
u		u		u		u		u
u		u		u		u		$\bar{u}$
d		d		d		d		

The total number of d quarks has remained the same, but we have increased the number of u quarks by one. We have also picked up a $\bar{u}$.

This sort of process in which a quark and an antiquark have been materialized is the opposite to the annihilation reactions that we have seen before. Previously we have noted that an e^+e^- pair can annihilate into energy. Now we have the materialization of a $q\bar{q}$ from energy. This process is evidently not completely arbitrary; there are physical laws that must be followed. These laws place limitations on what the reaction is able to do and they are best expressed in terms of the conservation of certain quantities. In turn, these conservation laws tell us about the fundamental forces that are at work—in this case the strong force.

7.2.1 Conservation of charge

All materialization and annihilation reactions must involve matter and antimatter. Conservation of electrical charge prevents the strong force from materializing quarks such as u and $\bar{d}$ from energy (the total charge is not zero). However, a $u\bar{c}$ materialization would conserve charge. Yet, such a materialization is not seen in strong force reactions. The properties of the strong force dictate that the quark and antiquarks involved must be of the same flavour. You can't materialize a u and a $\bar{c}$ from the strong force, but you can materialize a u and a $\bar{u}$.

We have seen charge conservation at work in lepton reactions. In this context it has implications for hadron reactions. For example:

$$p + p \rightarrow p + n + \pi^+ \tag{7.3}$$

is a possible reaction. At the quark level we have:

p	+	p	→	p	+	n	+	π^+
u		u		u		u		u
u		u		u		d		$\bar{d}$
d		d		d		d		

The number of u quarks in this reaction has not changed. The number of d quarks has increased by one and a $\bar{d}$ has also been produced. This would imply that the strong force has materialized a $d\bar{d}$ combination.

Notice that the d and the $\bar{d}$ *have not ended up in the same particle*. The π^+ contains one of the *original* u quarks plus the $\bar{d}$ produced by the strong force. The d that was also produced has changed places with the u in one of the protons (presumably the u that ends up in the π^+) turning that proton into a neutron[1].

Charge conservation is a very important rule as it allows us to say with certainty that many reactions that can be written down on paper will never be seen in experiments, such as:

$$p + p \nrightarrow p + p + \pi^+. \tag{7.4}$$

In this reaction the total electrical charge of all the particles on the right-hand side is greater by one unit than the total charge on the left. This is not a possible hadron reaction, and has never been seen in any experiment.

If reaction (7.4) were possible, then among the quarks we would see:

p	+	p	↛	p	+	p	+	π^+
u		u		u		u		u
u		u		u		u		$\bar{d}$
d		d		d		d		

the total number of u quarks has increased by one and we have also attempted to produce a $\bar{\text{d}}$. The only way this reaction could take place would be by the strong force materializing a $\text{u}\bar{\text{d}}$ combination, which has a total charge of $+1$. Conservation of charge is fundamental and so the strong force is unable to do this. The reaction cannot take place.

7.2.2 Conservation of baryon number

In the last chapter we noted that all quarks have $B = 1/3$ and that all antiquarks have $B = -1/3$. This clearly implies that any materialized $\text{q}\bar{\text{q}}$ pair must contribute $B = 0$ to the total baryon number in the reaction. The consequence of this is a curious restriction on the number of particles that can be produced. To take a simple case, we again react two protons together:

$$p + p \rightarrow p + p + \pi^0 + \pi^0. \tag{7.5}$$

This reaction conserves charge, so it will take place given sufficient energy. As protons consist of three quarks they have a total baryon number of $(+1/3 + 1/3 + 1/3) = 1$, so if we examine the baryon number for the reaction as a whole:

$$
\begin{array}{ccccccccccc}
& p & + & p & \rightarrow & p & + & p & + & \pi^0 & + & \pi^0 \\
B & 1 & + & 1 & = & 1 & + & 1 & + & 0 & + & 0
\end{array}
$$

showing that baryon number works like charge conservation—the total baryon number in the reaction does not charge. On the other hand, this reaction also conserves charge:

$$p + p \nrightarrow p + \pi^+ \tag{7.6}$$

but does *not* take place as it does not conserve baryon number. Baryon number conservation effectively counts the total number of quarks involved in the reaction. The quarks can appear either as qqq ($B = 1$), $\bar{\text{q}}\bar{\text{q}}\bar{\text{q}}$ ($B = -1$) or $\text{q}\bar{\text{q}}$ ($B = 0$); no other combinations are possible. The strong force can only materialize $\text{q}\bar{\text{q}}$'s out of energy, so the number of mesons produced in the reaction is limited only by the available energy, but the net number of baryons must remain fixed (i.e. the number of baryons minus the number of antibaryons).

Consider the following list of reactions:

$$p \;+\; p \;\;\not\to\;\; p \;+\; p \;+\; n \qquad\qquad (7.7)$$

$$p \;+\; n \;\;\not\to\;\; \pi^+ \;+\; \pi^0 \qquad\qquad (7.8)$$

$$p \;+\; \pi^+ \;\;\not\to\;\; p \;+\; p. \qquad\qquad (7.9)$$

In each case charge is conserved, but the reaction is blocked by baryon number. Of course protons and neutrons are not the only types of baryon:

$$p + p \to p + \Sigma^+ + K^0 \qquad\qquad (7.10)$$

$$p + p \to p + n + K^+ + \bar{K}^0 \qquad\qquad (7.11)$$

the rule works just as well for all baryons.

You may be wondering why it is not possible to split up the qqq combinations of baryons. After all, if one can materialize $q\bar{q}$ pairs, why can't the $\bar{q}$'s produced pair up with the quarks in the baryons to produce a flood of mesons?

In order to get the $\bar{q}$'s that we would need, the strong force would have to materialize them as $q\bar{q}$ combinations. Therefore the strong force has to produce three new quarks to go with the three antiquarks that we need to break up a baryon.

The three $\bar{q}$'s could then pair off with the quarks in the baryon, decomposing it into $q\bar{q}$ mesons. But that still leaves the three quarks that were produced at the same time as the $\bar{q}$'s! These three quarks can only bind together as a qqq combination—and so we are back to square one. Hence the number of qqq combinations must remain the same.

7.2.3 Conservation of flavour

The logic of baryon number conservation applies equally to the number of different types of quark in the reaction. When a $q\bar{q}$ is materialized by the strong force, the two must be of the same flavour. Consequently the materializations cannot alter the total number of quarks of each type in the reaction.

As we commented in the previous chapter, hadrons can be labelled with internal properties that indicate the number of quarks of a certain type within them. All hadrons that contain strange quarks are referred to as strange particles and have a value of strangeness (S). The total strangeness in any strong force reaction must remain the same. For example, if we collide two non-strange particles together, then we either do not create any strange particles in the reaction, or we create a pair with opposite strangeness:

$$p \; + \; p \; \rightarrow \; p \; + \; \Sigma^0 \; + \; K^+$$
$$S \quad 0 \; + \; 0 \; = \; 0 \; + \; -1 \; + \; 1 \tag{7.12}$$

$$p \; + \; \pi^+ \; \rightarrow \; \Sigma^+ \; + \; K^+$$
$$S \quad 0 \; + \; 0 \; = \; -1 \; + \; 1 \tag{7.13}$$

By an unfortunate historical accident the K^+ was chosen as the prototype strange particle and given $S = 1$. We now know that the K^+ is a $u\bar{s}$ combination and so the s quark has $S = -1$! Consequently the strangeness value of a hadron is equal to the number of strange quarks it contains in -1's.

This idea of conserving the individual flavours can be extended to the other heavy quarks as well: c, b and t. It has never proven useful to consider the separate conservation of u and d as they are very similar in mass.

It is important to remember that the weak force does not conserve particle types—it would allow reactions that do not conserve strangeness for example. The only situation in which this is likely to play a part is in the decay of particles (chapter 8); at other times the strong or electromagnetic forces dominate.

7.3   Using conservation laws

Conservation laws can be used to deduce information about new particles. In 1964 the following reaction was observed for the first time:

$$K^- + p \rightarrow K^0 + K^+ + X \tag{7.14}$$

Consider what we can deduce about the new particle X:

(1) X must have a negative charge—the total charge on the left-hand side is $(-1) + (+1) = 0$, as it stands the charge on the right-hand

side is $(0) + (+1) = 1$ so X must have charge -1 to make up the same total.

(2) X must be a baryon—the kaons on the right-hand side are mesons so X must have baryon number $B = 1$ in order to balance the $B = 1$ from the proton.

This tells us that X must be a qqq combination.

Now, the quark contents of the particles that we are familiar with are:

$$K^- + p \rightarrow K^0 + K^+ + X$$

K⁻	p	K⁰	K⁺	X
s	u	d	u	?
ū	u	s̄	s̄	?
	d			?

Note that the s quark in the initial K⁻ is unaccounted for. Quarks cannot just vanish. Hence X must contain this 'missing' s quark. One of the u quarks from the proton has gone to make up the K⁻ and one of the d quarks has gone to make up the K⁰, but what of the other u quark from the proton?

The clue to this lies in the two s̄ quarks that have appeared. If they have been materialized by the strong force (and where else could they come from?) then there must be two s quarks somewhere that were produced at the same time.

Where have they gone? Into X.

X is an sss combination. Two s quarks have come from the double s̄s materialization, and the other has come from the K⁻ that we started off with. So what of the missing u? Well there is one other particle unaccounted for, the ū from the K⁻. Evidently this ū has annihilated with the u from the proton back into energy! This reaction contains two materializations and an annihilation (dematerialization!). The reaction, then is:

$$K^- + p \rightarrow K^0 + K^+ + X$$

K⁻	p	K⁰	K⁺	X
s	u	d	u	s
ū	u	s̄	s̄	s
	d			s

This nice little exercise in logic has an extra point to it. This reaction was the first observation of the fabled Ω^- particle predicted by Murray Gell-Mann—the missing sss combination from the baryon decuplet (see chapter 6).

7.4 The physics of hadron reactions

In this section we are going to look a little deeper into the physics that lies behind the materialization and annihilation processes.

7.4.1 The field of the strong force

A force field is a volume of space inside which a force is exerted on an object. There is always another object that can be thought of as the source of the field. In the case of gravity one object, for example the earth, has a gravitational field that extends right out into space. Any object that has mass placed into this field experiences a force that attracts it towards the earth.

Despite their rather mysterious and intangible nature, gravitational fields are physically real objects. The field is a form of energy. As a mass falls towards the earth it collects energy from the field and converts this into kinetic energy. The total amount of energy that it is able to convert from the field into kinetic energy before it hits the ground is what we term the potential energy of the mass.

Electromagnetic fields act in a similar way. Any charged object is surrounded by an electric field and other charges placed within this field experience a force, the direction of which depends on the relative signs of the two charges (like charges repel, unlike charges attract). This field is slightly more tangible than the gravitational field, as it is possible to see what happens when the electric field is disturbed. If a charge is moved, then its field is disturbed. This disturbance travels through the field. If the charge is accelerating the disturbance travels as a wave—a light wave. Hence we can see what happens!

One common way of helping to visualize electrical fields is the *field line diagram*. These diagrams are a visual representation of an electrical field. Such a diagram is shown in figure 7.1.

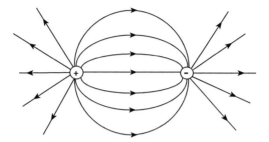

Figure 7.1. The electrical field between two attracting charges.

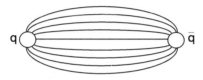

Figure 7.2. The strong field between a quark and an antiquark.

For quarks, the closest comparable situation to the electrostatic attraction of two charges is the attraction between a quark and an antiquark. However, the properties of the strong force are sufficiently distinct to make the field line diagram very different. Figure 7.2 is an attempt at a visual representation of the strong field between a quark and an antiquark.

The field lines are more concentrated between the quarks. Hence the energy of the field is concentrated between the particles in a narrow tube-like region called a *flux tube*.

If the force between two objects is attractive, then we must supply energy in order to separate them. If we want to pick up a rock and increase its distance from the centre of the earth, then we must supply energy. The energy that we supply is transformed into field energy. We have increased the energy in the field by separating the objects.

If we want to separate a quark from an antiquark we must supply energy. If we move the quark and antiquark further apart, then the flux tube connecting them gets longer. If the flux tube is getting longer there is more field, and hence more energy in the field.

In the case of the electrostatic or gravitational forces, the strength of the force decreases with distance. Specifically, if you double the distance between the charges (or masses) then the strength of the force drops to a quarter of what it was. This is known as the *inverse square law*.

If the force gets weaker with distance, then the energy needed to separate the objects decreases the further apart that they are. If two 1 coulomb charges are 1 metre apart, then to increase their separation by 1 centimetre takes 90 MJ of energy. However, if they start off 2 metres apart then to increase their separation by the same 1 centimetre only takes 23 MJ. In the case of the strong force the reverse is true. The strong force *gets stronger* the *further apart* the particles are.

If quarks are close up to each other (or q and q̄ in a meson) then there is virtually no force between them. If we try to separate them the strong force acts like an elastic band—the bigger the separation of the quarks, the greater the force pulling them back.

Separating quarks takes more and more energy the further apart they get, not less as in the electrostatic case. Consequently it is impossible to completely separate them: to do so would require an infinite amount of energy. This is the reason why the quarks can *never* be seen in isolation.

There is one problem with this. If the force between quarks increases with distance, how come there is no enormous force pulling all the quarks in my body towards one of the outer galaxies? Consider what happens when two baryons collide.

Figure 7.3(a) shows two protons moving towards each other at some speed. As a result of their collision, one of the quarks is given a tremendous thump by one of the quarks in the other proton. As a result it flies out of the proton taking a large amount of kinetic energy with it (figure 7.3(b))[2].

This quark will still be connected to the other quarks in the proton by a flux tube of strong field. As it moves away, this flux tube gets longer. In order to do this it must be gaining energy from somewhere—the obvious source being the kinetic energy of the ejected quark. As the quark moves away, its kinetic energy is being converted into strong field energy and hence into the lengthening flux tube—the quark slows down (which is

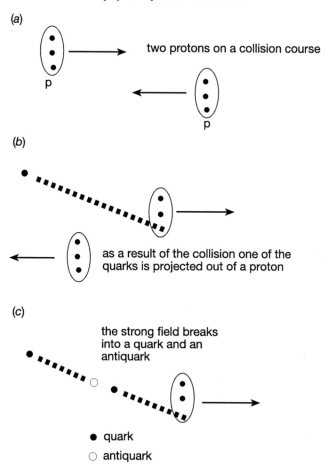

Figure 7.3. The collision of two protons leading to a reaction.

exactly what one would expect thinking of it in terms of an attractive force pulling the quark back into the proton).

Eventually, there is enough energy in the tube *to materialize a* $q\bar{q}$ *pair*. When this happens, the tube breaks into two smaller pieces (see figure 7.3(c)). The quark gets an antiquark on the other end of its piece of fractured tube, and the proton has a quark on the end of its tube. The ejected quark has turned into a meson, and the proton has gained a new quark to ensure that it remains a baryon, but not necessarily a proton. This is one physical process that lies behind hadron reactions.

How does this answer our original question about the long-range effects of the strong force?

If a quark in a distant galaxy were to interact with a quark in my body, then there would be a strong field between them. This field would be gigantic in its length, and hence would hold an enormous amount of energy. So much energy, in fact, that it would break up at myriad points along its length into quark–antiquark pairs. Hence the strong field cannot extend for very far because it becomes unstable and breaks up into mesons by materializing $q\bar{q}$ pairs. The strong force cannot extend for any great distance *precisely because its strength increases with distance*.

The flux tube mechanism can be used to explain the physical basis behind many hadron reactions. For example, if we look again at reaction (7.2):

$$p + p \rightarrow p + p + \pi^0. \tag{7.2}$$

The collision between the protons has caused one of the quarks to fly out, perhaps one of the u quarks. As it moves away the flux tube lengthens and then fractures into a $u\bar{u}$ pair, the $\bar{u}$ being on the part of the tube connected to the expelled u quark. This combination forms the π^0, while the u quark that was materialized combines with the other quarks to repair the fractured proton.

A full explanation of the physics of hadron reactions would take up more space and more technical detail than we can cover in a book of this sort. Indeed, although there are many models that are used to explain hadron reactions, a full theory has yet to be worked out. We are confident that all the physics involved is understandable in terms of the theory of the strong force, but the calculations required are beyond us at present. Simplifying assumptions have to be made. Different people argue for different assumptions, and so different models are used.

In any case, it seems clear from a study of the produced particles that slightly different mechanisms are at work depending on whether the particles are produced at the edge of the collision region (diffractive production, as it is called) or in the central 'fireball' region. The basic physics lies in the strong force materialization of $q\bar{q}$ pairs but the amount of energy and momentum involved, as well as which flavours are produced, are areas of detail that we are still working on.

7.5 Summary of chapter 7

- All reactions must conserve energy, momentum and electrical charge;
- hadron reactions conserve baryon number;
- baryon number conservation is a consequence of the limitations imposed on the materialization of new particles by the properties of the strong force;
- applying the conservation laws can help us to deduce the properties of new particles;
- fields are forms of energy that give rise to forces;
- the strong field between quarks is localized into flux tubes;
- as quarks get further apart the energy required to separate them increases;
- quarks can never be found in isolation;
- if there is enough energy in the flux tube it will fracture producing a new quark and antiquark at the ends of the pieces;
- this basic mechanism is responsible for hadron reactions.

Note

[1] Of course according to the quantum mechanics that we studied in chapter 3 it is impossible to tell which quark went where as they are identical. In a real calculation of this process all possibilities would have to be taken into account.

[2] This is an oversimplified model. In reality more than one quark will be jolted and there will be flux tubes connecting all the quarks together.

Chapter 8

Particle decays

In the last chapter we saw that hadron reactions are determined by the actions of the strong force. In this chapter we shall see that the weak and electromagnetic forces also have roles to play in particle decays. By studying the decays of particles we shall learn more about these forces.

Particle decay is an example of nature's tendency to reduce the amount of energy localized in an object. Nature prefers energy to be spread about as much as possible. When a particle decays the same amount of energy is shared out among a greater number of objects.

This tendency is reflected in other processes that are similar to particle decay. Radioactivity is an obvious example, but the emission of light by atoms also has features in common with decay processes.

8.1 The emission of light by atoms

When an electron is contained within an atom it is forced to adopt an exact energy value. The value is referred to as an energy level. This is a consequence of the closed nature of paths within an atom. The electron's Lagrangian restricts the energy values to ones that allow the phase to return to its initial value as you follow it round a closed path. There are many energy levels to choose from (subject to a strict set of rules), but the electron can only choose an exact level—it cannot take an energy value that falls between levels.

Sometimes an electron may be in a higher energy level than normal. When this happens the atom is said to be in an *excited state*. Atoms can be put into excited states by absorbing energy from outside, e.g. if the material is heated. The electron does not stay in this higher energy level if there is a space for it in a lower level—it 'falls' into the lower level and in so doing reduces its energy. Energy is, of course, conserved so some other object must gain energy if the electron loses it. The energy is emitted from the atom in the form of a burst of electromagnetic radiation called a photon—light is emitted.

8.2 Baryon decay

8.2.1 Electromagnetic decays

We know from chapter 6 that sometimes two distinct hadrons have the same quarks inside them and differ only in the energy levels that the quarks are sitting in.

Does this mean that we can get transitions in the energy levels of the quarks as we can with the electrons in atoms? What would we observe if such an event were to take place?

The previous discussion would suggest that we would see the emission of a photon as the drop in energy level took place. How would we go about looking for such an event?

For such an event to take place the quarks within the hadron would have to change state. If the state of the quarks is related to what type of hadron it is, as we have just recalled, then the process would result in the hadron changing into another hadron. For the moment we choose to look for such a possibility amongst the baryons.

There is a classic example that we can study:

$$\Sigma^0 \rightarrow \Lambda + \gamma. \tag{8.1}$$

This is an example of a decay equation. The object on the left-hand side of the equation was present before the decay and is replaced by the objects on the right-hand side of the equation after the decay.

Particle decays

We should look at the process with more care. At the quark level:

$$
\begin{array}{ccc}
\Sigma^0 & \rightarrow & \Lambda + \gamma \\
u & & u \\
d & & d \\
s & & s \\
\end{array}
$$

Mass   1.192   1.115
(GeV/c^2)

The quark contents of the two hadrons are identical. The extra mass of the Σ^0 comes from the quarks being in higher energy levels. Evidently, the decay takes place because the quarks have dropped into a lower energy state and the excess energy has been emitted in the form of a photon. This will happen spontaneously.

This decay of the Σ^0 has been closely studied, and it has been found that, on average, Σ^0 particles will decay in this way within 2.9×10^{-10} seconds of their creation in a hadron reaction. We call this an *electromagnetic* decay as a photon is emitted in the process.

If we want to find other examples then we need a hadron that has a lighter mass double into which it can decay. An obvious place to look would be amongst the Δ family of particles—the Δ^+ has the same quark content as that of a proton and the Δ^0 that of the neutron. Perhaps these particles have electromagnetic decays.

If you look them up in a data table, then you will find that about 0.6% of deltas do decay in the way that we have suggested:

$$\Delta^+ \rightarrow p + \gamma \tag{8.2}$$
$$\Delta^0 \rightarrow n + \gamma. \tag{8.3}$$

However, in most cases the deltas decay by an alternative mechanism:

$$\Delta^+ \rightarrow p + \pi^0 \tag{8.4}$$
$$\Delta^0 \rightarrow n + \pi^0. \tag{8.5}$$

Both these decays take place extremely rapidly. Typically a delta will decay some 10^{-25} seconds after it is formed[1]. Decays (8.4) and (8.5) clearly represent better options that the deltas nearly always take.

8.2.2 Strong decays

Let's look at these decays at the quark level:

$$\Delta^+ \rightarrow p + \pi^0$$

u		u	u
u		u	ū
d		d	

and

$$\Delta^0 \rightarrow n + \pi^0$$

u		u	u
d		d	ū
d		d	

As there is an obvious similarity between the two decays, we can concentrate on the first without neglecting any important physics.

Notice that the right-hand side of the quark list looks just like what we might expect in a hadron reaction *producing* the π^0. We saw something very similar when we looked at the reaction $p + p \rightarrow p + p + \pi^0$ in section 7.2. In that reaction we discovered that the strong field had been disturbed by quarks colliding. The field then used the energy it gained to materialize a quark/antiquark pair, which went on to form the pion.

If something similar is happening in (8.4) the strong field inside the Δ^+ must have enough energy to materialize the uū pair. We can check that by looking at the masses of the Δ^+ and the proton:

$$p \text{ mass} = 0.938 \text{ GeV}/c^2$$
$$\pi^0 \text{ mass} = 0.135 \text{ GeV}/c^2$$
$$\overline{\text{total} = 1.073 \text{ GeV}/c^2}$$
$$\Delta^+ \text{ mass} = 1.23 \text{ GeV}/c^2.$$

On this basis there looks to be more than enough energy! The total mass of the proton and a π^0 is much less than that of a Δ^+. This is an example of a *strong decay*.

Inside a hadron, the quarks are bound to each other by the strong force. Hence the quarks within the hadron are wading about in a strong field. The strong force is very much stronger than the electromagnetic forces

Particle decays

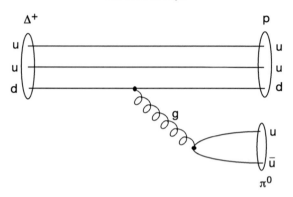

Figure 8.1. The Δ^+ decay drawing.

between the quarks and so it is more significant in determining their energy levels. When the quarks lose energy a disturbance is set up in the strong field—the energy is localized in a certain part of the field (I visualize it to be like a lump in some custard, the custard representing the strong field). This disturbance is emitted in a similar way to the photon in the electromagnetic case. A moving packet of disturbance in the strong field is called a *gluon* (pronounced 'glue-on'). Gluons are to the strong force what photons are to the electromagnetic force.

The analogy between photons/electromagnetic field and gluons/strong field is a good one and has proven very useful in working out the theory of the strong force.

We conclude that the quarks in the Δ^+ are able to switch energy levels by emitting a gluon. However, the strong force has a trick up its sleeve. Out of the energy of the gluon materializes a quark and an antiquark, in this case either a $u\bar{u}$ or a $d\bar{d}$ that form the π^0.

Physicists like to draw sketch diagrams such as figure 8.1 to illustrate such processes.

In this drawing the complete lines represent quarks that are 'moving' through the diagram, in the sense that the left-hand side of the diagram represents the initial state (the Δ^+ in this case) and the right-hand side represents the final state. The loops at each end show that the quarks are

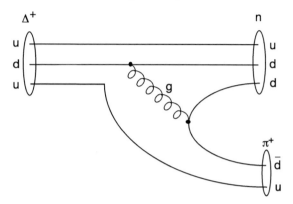

Figure 8.2. An alternative Δ^+ decay.

bound into hadrons, and the specific hadron is labelled above the loop. The curly line is the gluon. As you can see, the diagram shows a quark in the Δ^+ emitting a gluon which then materializes into a $u\bar{u}$ combination that forms a pion.

Be careful not to read too much into this diagram. The lines are *not* supposed to represent definite paths of the particles. It is just a sketch to help visualize what is happening.

There is another way in which the Δ^+ can decay via the strong force:

$$\Delta^+ \rightarrow n + \pi^+. \tag{8.6}$$

This is a slightly more complicated case. We can explain it by presuming that the gluon materializes into a $d\bar{d}$ combination, but that they do not go on to bind to each other; instead the d joins with the u and one other d from the initial Δ^+ to form the neutron, and the $\bar{d}$ joins with the leftover u from the Δ^+ to form the π^+. One of our diagrams (figure 8.2) will help in visualizing this.

The rest of the delta family of hadrons can also decay:

$$\Delta^{++} \rightarrow p + \pi^+ \tag{8.7}$$
$$\Delta^- \rightarrow n + \pi^- \tag{8.8}$$

with similar lifetimes to the decays that we have been discussing.

Table 8.1. The baryon decuplet.

Particle family							Average mass (GeV/c^2)
Δ^-		Δ^0		Δ^+		Δ^{++}	1.23
	Σ^{*+}		Σ^{*0}		Σ^{*-}		1.39
		Ξ^{*0}		Ξ^{*-}			1.53
			Ω^-				1.67

Returning to the Σ^0 decay, it is interesting to see why the decay cannot proceed via the strong force. After all, there does appear to be a process that conserves both charge and strangeness open to it:

$$\Sigma^0 \rightarrow \Lambda + \pi^0. \tag{8.9}$$

In order to see the answer we must consider the masses of the particles concerned.

$$\Lambda \text{ mass} = 1.115 \text{ GeV}/c^2$$
$$\pi \text{ mass} = 0.135 \text{ GeV}/c^2$$
$$\text{total} = 1.250 \text{ GeV}/c^2$$
$$\Sigma^0 \text{ mass} = 1.193 \text{ GeV}/c^2.$$

The total mass of a Λ and a π^0 is *greater* than the mass of a Σ^0. The strong decay is blocked as there is not enough excess energy in the Σ^0 to create the pion out of the strong field, hence the decay proceeds via the electromagnetic force.

We can divide the particles that we discussed in chapter 6 into groups depending on the manner of their decay. We have two sets of baryons to deal with (see tables 8.1 and 8.2), the decuplet and the octet. The octet is the set of lower mass baryons some of which are reflected in higher mass versions that lie in the decuplet.

We have already dealt with the delta decays, but now we see that the high mass particles in the decuplet can decay into low mass versions in

Table 8.2. The baryon octet.

Particle family			Average mass (GeV/c^2)
	p	n	0.939
Σ^+	Σ^0/Λ	Σ^-	1.19/1.12
	Ξ^0	Ξ^-	1.32

the octet. For example:

$$\Sigma^{*+} \rightarrow \Sigma^+ + \pi^0 \qquad (8.10)$$

$$\Sigma^{*+} \rightarrow \Lambda + \pi^+ \qquad (8.11)$$

$$\Sigma^{*0} \rightarrow \Sigma^0 + \pi^0 \qquad (8.12)$$

are all possible decays. Note that the Σ^{*0} has a strong decay open to it which it will take in preference to the electromagnetic one (in this case there is a big enough mass difference). I am sure that you can make up other possible decays for the particles in the decuplet. Just to emphasize the point here are a few more:

$$\Xi^{*-} \rightarrow \Xi^- + \pi^0 \qquad (8.13)$$

$$\Xi^{*0} \rightarrow \Xi^0 + \pi^0 \qquad (8.14)$$

$$\Xi^{*0} \rightarrow \Xi^- + \pi^+. \qquad (8.15)$$

However, when we come to the Ω^- we hit a problem. There is no member of the octet with the same quark content (sss) and so there is no strong or electromagnetic decay open to it. The Δ^{++} (uuu) and the Δ^- (ddd) can decay by the strong force using a mechanism (figure 8.3) similar to one we have drawn before and it is possible to draw a similar diagram for the Ω^- (figure 8.4).

However, this decay is blocked as there is not enough mass:

$$\Omega^- \rightarrow \Xi^- + \bar{K}^0$$

$$\Xi^- \text{ mass } = 1.197 \text{ GeV}/c^2$$

$$\bar{K}^0 \text{ mass } = 0.498 \text{ GeV}/c^2$$

$$\text{total } = 1.695 \text{ GeV}/c^2$$

$$\Omega^- \text{ mass } = 1.672 \text{ GeV}/c^2.$$

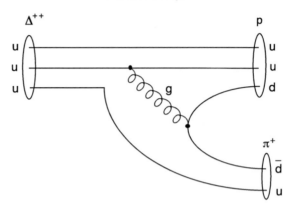

Figure 8.3. The Δ^{++} strong decay.

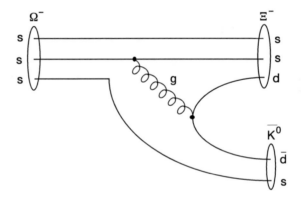

Figure 8.4. A possible Ω^- decay.

Gell-Mann was aware of this when he predicted the existence of the Ω^- and that it would have a lifetime longer than those of the other members of the decuplet. As we have seen, the strong decays typically take place in about 10^{-25} seconds, whereas the Ω^- typically takes 1.3×10^{-10} seconds to decay.

8.2.3 Weak decays

Before reading this section it would be wise to review the section of chapter 7 called 'the field of the strong force'. We are going to use the ideas of a field quite a lot in this section, so you ought to be sure that you are reasonably happy with what we have said so far on the subject before carrying on.

At the end of the last section we left open the question of how the Ω^- particle was able to decay as the strong and electromagnetic channels were blocked to it. The same is true for the majority of particles in the baryon octet. The Σ^0 can decay electromagnetically into the Λ but there are no light mass versions of the Σ^+ for example (the octet is after all the lowest mass versions of these quark contents), so how can these particles decay?

One of the characteristics of the weak force is that it allows quarks to change flavour, and that is what is needed here. If for example the Σ^+ were to decay, the only particles lighter than it would be the proton and neutron, neither of which contains a strange quark. The strange quark has to be removed in order for the Σ^+ to decay.

The Σ^+ has several possible decays open to it, but the most common ones are:

$$\Sigma^+ \rightarrow p + \pi^0 \tag{8.16}$$
$$\Sigma^+ \rightarrow n + \pi^+. \tag{8.17}$$

Let us look at these at the quark level to see if we can deduce what is going on:

$$\Sigma^+ \rightarrow p + \pi^0$$

u	u	u
u	u	$\bar{u}$
s	d	

It is obvious from this that two things have happened: the s quark has disappeared and we have picked up u, $\bar{u}$ and d quarks instead. Evidently the s has been turned into one of these other quarks and two more have appeared—probably out of the field as in the case for strong decays.

If this decay is to follow the pattern that we have already established for the strong and electromagnetic decays then there must be some sort of transition from a high energy state into a low energy state coupled by the emission of some disturbance to the field. Obviously in this case the transition also entails turning an s quark into some other flavour—either

Table 8.3. The action of the weak force on the quarks and leptons.

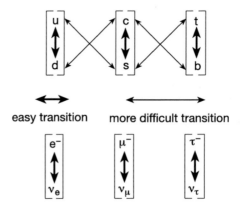

u, d or ū according to the quarks that turn up on the right-hand side of the decay equation.

Years of work and study of the processes of weak decays and interactions have established that the weak force must follow a set pattern with regard to the quarks. The evidence for this pattern is compelling but complicated, so I am going to have to ask that you accept it as one of the rules. The weak force acts on the quarks in the manner outlined in table 8.3.

Table 8.3 is telling us is that it is possible for the weak force to change the s quark into a u quark. This is obviously a beneficial transition to take place within the Σ^+ as the u quark is less massive, and therefore the energy of the hadron will be reduced. This is not quite the same as a quark dropping energy levels, in a sense it is more fundamental than that. In this case the quark is changing into a different particle and in the process giving up some energy to the weak field.

A quark has an electrical charge and consequently is always surrounded by an electromagnetic field. Similarly, a quark is always surrounded by strong and weak fields.

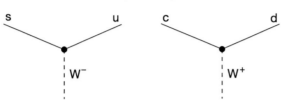

Figure 8.5. Some of the weak transitions of quarks.

Within the Σ^+ the s changes into a u producing a disturbance in the weak field. This disturbance is similar to a gluon or a photon in some ways, but there is a very important difference: it is electrically charged[2].

Just putting to one side for the moment what that actually means, let us see *why* it must be so.

If an s quark ($-1/3$) can change into a u quark ($+2/3$) by giving energy over to the weak field, then it must give over charge as well. If we say that the field gains a charge Q, then conservation of charge suggests:

$$\text{initial charge (s quark)} = -1/3$$
$$\text{final charge (u quark)} = +2/3 + Q$$

hence

$$-1/3 = +2/3 + Q$$
$$\therefore \quad Q = -1.$$

When a quark changes state in a weak field it gives up both energy and charge to the field, and the object that is produced is called a 'W'.

Actually, there are two types of W, the W^+ and the W^-. It is clear that when the s changes to a u quark it must emit a W^- in order to balance the charge change. Similarly, if a c were to change into a d, then it would have to emit a W^+.

Particle physicists like to draw sketches (as in figure 8.5) to help them remember this.

If you find it difficult to remember then you can always work it out by balancing the charges (that's what I do—I can never remember it!).

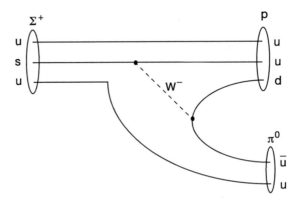

Figure 8.6. The Σ^+ decay.

So far we have said that the s turns into a u by emitting a W^-, but what then happened to this? In the case of the strong decay the gluon materialized a $q\bar{q}$ pair. The same thing happens to the W^-, but true to the nature of the weak force, the quark and antiquark do not have the same flavour. In this case, the W^- materializes into a $d\bar{u}$ pair. The gluon is not electrically charged, so it is only able to materialize quark/antiquark pairs of the same flavour.

In order to summarize all this, we will draw one of our sketches, see figure 8.6.

8.2.4 Neutron decay

The neutron is an unstable particle when it is in isolation. Inside the nucleus, the effect of the other protons and neutrons is to stabilize the individuals so that they can last indefinitely. If this stabilizing effect is not quite complete, then the result is a form of β radioactivity.

The decay process is:

$$n \rightarrow p + e^- + \bar{\nu}_e \tag{8.18}$$

or at the quark level

$$d \rightarrow u + e^- + \bar{\nu}_e. \tag{8.19}$$

The existence of this process was well known in the 1930s, but it presented a profound difficulty to the physicists of the time. The

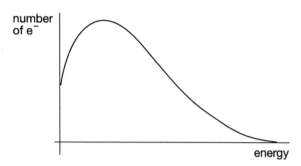

Figure 8.7. The kinetic energy distribution of e⁻ produced in nuclear β decay.

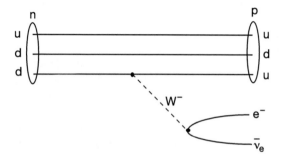

Figure 8.8. Neutron decay.

existence of the $\bar{\nu}_e$ was unsuspected. Only the emitted e⁻ could be detected. Measurements of the e⁻ energy showed that it was produced with a wide range of energy values up to some maximum (figure 8.7).

This seemed to violate the law of conservation of energy which would suggest that as there were only two objects involved after the decay (the proton and the electron) the split of the available energy between the two of them should be fixed—hence the e⁻ should always have the same energy. Wolfgang Pauli suggested that the puzzle could be solved if another neutral particle was present after the decay making the split of energy a three-way process and hence not fixed. It was not until 1953 that the neutrino was discovered (see page 77).

Neutron decay (figure 8.8) is a typical weak decay which is even slower than normal as the proton and neutron have only a very small mass difference between them.

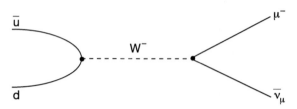

Figure 8.9. The π^- decay.

8.3 Meson decays

As the W's can materialize into q$\bar{\text{q}}$ pairs of different flavour, it should not surprise us to discover that the weak field can also be involved in the annihilation of quarks of different flavours.

Direct evidence of this process can be seen in the decay of the lightest mesons. The simplest family of mesons is the pion family π^+, π^- and π^0.

The π^+ and the π^- are antiparticles of each other containing u$\bar{\text{d}}$ and d$\bar{\text{u}}$ quark combinations respectively. Conservation of electrical charge and flavour forbids the pions to decay via the strong or electromagnetic forces, so a weak decay is the only avenue open to them.

We already know from the Σ^+ decay that the W$^-$ is capable of materializing into a d$\bar{\text{u}}$ quark combination. In the case of the π^- the same quark combination is capable of mutual annihilation by turning into a W$^-$.

As the d and $\bar{\text{u}}$ quarks are amongst the lightest, there is no advantage in the π^- decaying into a W$^-$ which then materializes quarks again. The only other possibility is for the W$^-$ to turn into a lepton pair:

$$\pi^- \rightarrow \mu^- + \bar{\nu}_\mu. \tag{8.20}$$

We can sketch this process as in figure 8.9.

This is just as much an annihilation reaction as the e$^+$e$^-$ reactions and the q$\bar{\text{q}} \rightarrow$ g reactions that we have studied. In this case it is the weak field which is excited by the annihilation rather than the electromagnetic or the strong.

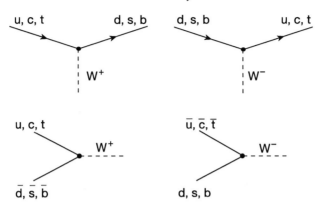

Figure 8.10. The W couplings to the quarks and antiquarks.

Notice that the decay does conserve both baryon number ($B = 0$ initially) and lepton number.

The weak field is therefore capable of a complex set of reactions. It can either cause flavour transformations, annihilations or materializations. It is possible to summarize all the possibilities in a set of sketches (see figure 8.10).

These sketches show the ways in which quarks can 'couple' to the weak field. Notice that the lower two sketches in the set indicate that the $q\bar{q}$ pairs listed can annihilate into a W, or in reverse the W can materialize into the pair. Remembering all these couplings is not quite as daunting as it seems. Once you have mastered the quark generations simple conservation of charge enables one to work out the possibilities.

The third member of the pion family, the π^0 also decays, but as it has the quark content $u\bar{u}/d\bar{d}$ it can decay electromagnetically—matter and antimatter annihilating into photons (see figure 8.11):

$$\pi^0 \rightarrow \gamma + \gamma. \tag{8.21}$$

Notice that conservation of energy and momentum forbids the π^0 from decaying into a single photon. Inside the pion a continual conversion takes place involving the exchange of two gluons (see figure 8.12), which is why the π^0 is both a $u\bar{u}$ and a $d\bar{d}$ combination.

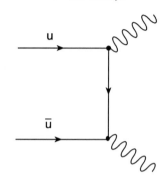

Figure 8.11. The π^0 decay.

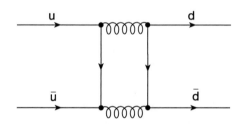

Figure 8.12. Inside a π^0.

8.4 Strangeness

In the early days of hadron reactions, it became clear that there was a class of particle that could be produced quite easily in the reactions, were quite massive and yet took comparatively long times to decay. The fact that they were produced quite easily suggested that they were being created by the action of the strong force, making them hadrons. If this was true, then it was puzzling that they did not seem to decay via the strong force as well. This was clearly not the case as they had lifetimes typically of the order of 10^{-10} s.

The other conspicuous feature was that these particles were only ever created in pairs e.g.:

$$p + p \rightarrow p + \Sigma^+ + K^0 \tag{8.22}$$

but never singly,

$$p + p \nrightarrow p + \Sigma^+ + \pi^0. \tag{8.23}$$

This feature led Gell-Mann, Nakano and Nishijima in 1953 to suggest that the whole puzzle could be solved by assuming the existence of a new internal property of hadrons. They named this property *strangeness* (after the odd behaviour of the particles) and suggested that it must be conserved in strong interactions. This explained why the particles were produced in pairs. In reaction (8.22) the protons both have strangeness $S = 0$, so the initial total is zero. After the reaction, the Σ^+ has $S = -1$ and the K^0 has $S = +1$ making the total zero again.

When the particles come to decay, they can decay into another strange particle if there is a less massive one in existence e.g.:

$$\Sigma^{*+} \rightarrow \Sigma^+ + \pi^0 \tag{8.10}$$

which would be a fast strong decay. However, if there was no lighter strange particle available then the strong force could not be used as the decay would have to involve a loss of strangeness;

$$\Sigma^+ \quad \nrightarrow \quad p \quad + \quad \pi^0$$
$$S \quad -1 \quad \neq \quad 0 \quad + \quad 0.$$

The suggestion was that the weak force did not conserve strangeness and hence would be the force involved in the decay. Hence the lifetime would be much longer. This suggestion of a new internal property put Murray Gell-Mann and others on the road to the classification of hadrons that we discussed in chapter 6 (the eightfold way) and eventually to the discovery of quarks.

8.5 Lepton decays

Everything that we have said about the decay of hadrons can be applied to the decay of leptons. We already know from chapter 4 that some leptons can decay. For example:

$$\mu^- \rightarrow e^- + \nu_\mu + \bar{\nu}_e. \tag{5.2}$$

This decay proceeds via the emission of a W particle by the μ^- which changes it into a ν_μ (see figure 8.13).

Tau decay takes place in a similar manner, but electrons are prevented from decaying by conservation of lepton number.

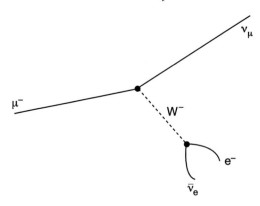

Figure 8.13. The decay of the μ^-.

8.6 Summary of chapter 8

- Particles decay in order to rid themselves of excess energy;
- a simple electromagnetic decay is caused by quarks dropping into lower energy levels and emitting photons;
- an electromagnetic decay has a lifetime typically of the order of 10^{-20} seconds;
- strong decays take place when quarks drop into lower energy levels and emit gluons;
- the emitted gluon then materializes into a $q\bar{q}$ pair;
- strong decays have lifetimes typically of the order of 10^{-25} seconds;
- a weak decay takes place when a quark/lepton changes into another quark/lepton by emitting a W^+ or a W^-;
- the W's then materialize into a $q\bar{q}$ pair or a lepton and an antilepton (conserving lepton number);
- weak decays have lifetimes typically of the order of 10^{-10} seconds.

Note

[1] This is such a short time that most people refuse to dignify them with the name *particle*, preferring to call them *resonances*. I will continue to call them particles so as to save introducing a new term.

[2] There is also a neutral weak force particle called the Z^0, see section 11.2.2.

Chapter 9

The evidence for quarks

The evidence for the existence of quarks has built up over
the last thirty years. Physicists have taken some time to be
convinced by the various threads of evidence. In this chapter
we shall look at the various arguments that have been put
forward to justify the belief in quarks.

9.1   The theoretical idea

In 1964 Gell-Mann and Zweig independently came up with the idea that
the 'eightfold way' pattern of arranging hadrons (see chapter 6) could be
explained by taking three basic sets of properties and combining them in
various patterns[1]. As a mathematical structure the idea was very pleasing
but there was no direct experimental evidence for it and it did raise some
disturbing theoretical problems, so it was not widely accepted. However,
the search for elegance is a powerful motivator in physics and this was
certainly an elegant idea, so it did attract some attention.

9.2   Deep inelastic scattering

If physicists required some more direct evidence for the existence of
quarks then it was not long in coming. The first real evidence came as
a result of the analysis of data taken by an experiment at the Stanford
Linear Accelerator Centre (SLAC) in the late 1960s. The leaders of
this experiment, Friedman, Taylor and Kendal, were awarded the 1990
Nobel Prize in Physics. The Stanford experiment was, in essence,

a generalization of a famous experiment performed by Geiger and Marsden in 1909. As the first person to produce a theoretical explanation for their results was Ernest Rutherford, the physics involved has become known as 'Rutherford scattering'.

9.2.1   Rutherford scattering

When this experiment was first performed physicists were using a model of the structure of the atom that had been proposed by J J Thomson in 1903. Thomson suggested that the negatively charged electrons, that were known to be present inside atoms, were embedded into a positively charged material that made up the bulk of the atom (protons had not been discovered). The atom was a solid ball into which the electrons were stuck, rather like the plums in a pudding. Thomson's idea became known as the 'plum pudding' model of the atom.

As the nature of the positive material was unknown, Rutherford was keen to find out whether the positive 'stuff' was uniformly spread throughout the volume of the atom, or distributed in some other fashion. Earlier experiments had shown him that α particles could be deflected through large angles when passed through thin sheets of mica. However, Rutherford was unable to produce similarly large deflections by applying very strong electrical or magnetic fields directly to α particles. Clearly the forces at work within the mica sheets were far greater than those that Rutherford could produce in the lab.

In 1909 Rutherford suggested to Ernest Marsden, a student of Rutherford's collaborator Hans Geiger[2], that he use α particles from a naturally radioactive isotope to try to determine how big a deflection could be produced. Rutherford was hoping to find the distribution of positive charge within the atom by studying the ways in which the positively charged α particles were deflected by electrostatic repulsion.

Marsden elected to use a thin foil of gold as the target for the α particles. Gold is a relatively heavy atom and the metal can easily be formed into thin sheets. The thinner the target, the smaller the number of atoms that the α particles would collide with as they passed through, making the collisions easier to study. (α particles do not penetrate very far through matter, so the foil had to be thin for another reason—the α particles would never make it through to the other side if it was any thicker!)

The result of Marsden's experiment came as a total surprise to Rutherford and Geiger. Marsden discovered that most of the α particles passed through the foil and were hardly deflected. However, approximately 1 in 20 000 were deflected to such an extent that they reversed course and bounced back in the direction that they had come from. Rutherford described this as being like firing a 15 inch shell at a piece of tissue paper and watching it bounce off.

Although only a small proportion of the α particles were bounced back, calculations convinced Rutherford that positive material distributed throughout the atom could not be concentrated enough to exert such a large force on an α particle. This puzzle worried Rutherford for at least a year until in 1910, with the aid of a simple calculation, he saw the significance of Marsden's results.

He reasoned that if an α was to be turned right back, then it must come to rest at some distance from the centre of the atom and then be pushed back. If it came to rest then all its kinetic energy must have turned into electrostatic potential energy at that distance. Knowing the amount of kinetic energy the α had to start with it was a simple matter to calculate the distance. The answer astounded him. The α particles that were completely reversed had to come within 10^{-15} m of the *total* positive charge of the atom. As the diameter of a gold atom was known to be of the order of 10^{-10} m, this implied that *all* the positive charge was concentrated in a region at the centre of the atom some hundred thousand times smaller than the atom itself. Thomson's model of the atom could not have been more wrong[3].

Rutherford used this result as the justification for his nuclear model of the atom. He suggested that all the positive charge is concentrated in a very small region, known as the nucleus, in the centre of the atom. The electrons were in orbit round this region, rather than embedded within it, in a similar way to that in which the planets orbit the sun. We now appreciate that this crude model is inadequate in many important ways[4], but the essential feature of the nucleus is retained.

Figure 9.1 contrasts Thomson's and Rutherford's atomic models. It was not long before Rutherford's model was refined by the discovery of the proton and neutron inside the nucleus. This, in essence, is the model that we still use today.

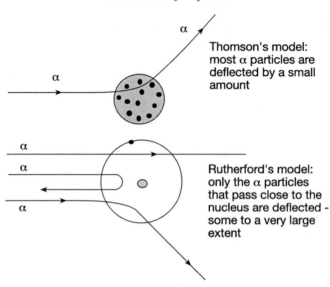

Figure 9.1. Contrasting models of the atom.

9.2.2   The SLAC experiment

The SLAC experiment worked on the same basic theme as Rutherford scattering: using a particle to probe the structure of an object. In this case, however, the probe particles were electrons accelerated to very high energies and the object was the proton.

Electrons are significantly easier to use than α particles. They are easily produced in large numbers; they have low mass, and so can be accelerated comparatively easily, and they do not feel the strong interaction.

The last point is significant. The Stanford physicists intended to fire high energy electrons at protons and to measure the extent to which they were deflected to see how the electrical charge is distributed throughout the volume of the proton. If they tried to use a hadron as the probe, then it would react with the proton via the strong interaction. Such reactions are much harder to analyse than those caused by the electromagnetic force. In any case, the strong force does not depend on the electrical charge of the objects concerned, so the resulting reactions would not provide any direct information about the charge distribution within the proton.

What they found took them by surprise.

At low energy the electrons deflected slightly and the protons recoiled. As they increased the energy of the electrons a threshold energy was reached beyond which the electrons were deflected through much greater angles. At the same time the protons started to fragment into a shower of particles rather than being deflected themselves. They had started to see evidence for charged quarks inside the proton.

9.2.3 Deep inelastic scattering

At the time, the results of the SLAC experiments were a great puzzle— the quark theory was not well known. The full theory that explained the results was developed by Feynman and Bjorken in 1968. Feynman named the small charged objects inside the proton 'partons'. However, it soon became clear that Feynman's partons were the same as Gell-Mann's quarks.

(Some people maintain a distinction by using the name parton to refer to *both* the quarks and the gluons inside protons and neutrons.)

The effect of a high speed charged object passing close to the body of another charged particle, such as a proton, is to produce a severe disruption in the electrostatic field that connects the particles. The exact nature of this disruption depends on the motion of each of the charges involved. At a crude level of approximation we can think of this disruption as being caused by the passage of a photon from the electron to the proton. We can draw one of our sketches to represent this process (see figure 9.2).

It is important to have some physical picture of what is happening in this process. Many authors have suggested various ways of looking at this, none of which seem entirely satisfactory to me. I wish to base my account on the very physical picture used by Feynman.

The emission of a photon by an electron is not independent of its subsequent absorption. Diagrams like figure 9.2 encourage us to think of

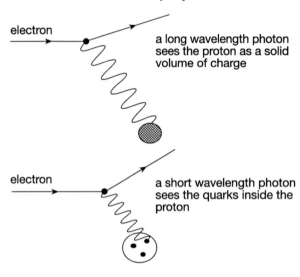

Figure 9.2. The interaction between a photon and a proton.

the photon as a particle moving through empty space (as symbolized by the wavy line). In truth it is more like the motion of a ruck in a carpet that is being pulled tight. The carpet is important in determining the motion of the ruck. Although we do not draw the electromagnetic field on our diagram, it is there and it connects the charged objects (it is the 'carpet'). Whether it is a high or low energy photon that is passed depends on the motion of *both* charges and the consequent disturbance of the *total* field.

A low energy electron is moving comparatively slowly and so takes some time to pass by the proton. In this time the quarks within the proton, which are moving at a very high speed, cover considerable distances although they remain localized within the proton. Hence the electron 'sees' an average electromagnetic field that is smeared out over the volume of the proton due to the motion of the quarks. (Do not forget that the charge of the proton is simply the sum of the charges of the quarks inside.) This 'smeared' field is of relatively low intensity and so the interaction only triggers a low energy photon from the electron. If the electron emits a low energy photon then it is hardly deflected from its path and the photon, which has a long wavelength, is absorbed by the whole proton.

On the other hand, a high energy electron passes the proton very quickly. In the time it takes to pass the proton the quarks hardly move[5] and so the electromagnetic field has several localized intense regions (corresponding to the positions of the quarks). An electron passing close to one of those regions will cause a severe localized disturbance in the field—a high energy photon is emitted and then absorbed by the quark.

Imagine a set of helicopter blades. An object passing by the blades slowly would see them as an apparently solid disc. A quick object would see individual blades as they do not have the chance to move very far while the object passes.

If the electron emits a high energy photon (short wavelength) it is deflected through a large angle. The photon is absorbed by an individual quark within the proton. As a result of this the quark is kicked out of the proton as well. The quark will move away from the proton, stretching the strong field until it fragments into a shower of hadrons. The result is a deflected electron and a shower of hadronic particles.

When the experimental results show that the electrons are being deflected by large angles and the proton is fragmenting into a shower of hadrons, then we can deduce that the electron is interacting directly with the quarks within the proton.

The SLAC experiment ran for many years and was refined by using other particles, such as neutrinos and muons as probes. The SLAC people were able to establish that there are three quarks inside the proton and that their charges correspond to those suggested by Gell-Mann and Zweig. Deep inelastic scattering forms the cornerstone of our physical evidence for the existence of quarks.

9.3 Jets

The third major piece of evidence to support our belief in the existence of quarks comes from the study of e^+e^- annihilation reactions. We first mentioned annihilation reactions in chapter 5 when we saw that they could be used to produce new flavours of leptons. The basic reaction takes the form

$$e^+ + e^- \rightarrow \mu^+ + \mu^- \tag{9.1}$$

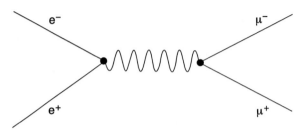

Figure 9.3. A typical annihilation reaction.

a reaction that becomes possible once there is enough energy in the incoming electron–positron pair to materialize the muon and the antimuon. Since then we have looked at similar processes, hence we recognize that the electromagnetic force is taking energy from the annihilation and is then materializing new particles from that energy. We can even draw a sketch diagram to illustrate the process (see figure 9.3) in which the photon represents the energy flowing into and out of the electromagnetic field.

Evidently the electromagnetic force has a free hand when it comes to materializing the new particles. Baryon number, lepton number and electrical charge are all zero initially, so as long as opposites are created, then anything can come out the other end. Specifically the reaction:

$$e^+ + e^- \rightarrow q + \bar{q} \tag{9.2}$$

is perfectly possible, provided the q and the $\bar{q}$ are the same flavour (i.e. $u\bar{u}$, $d\bar{d}$, etc).

In the case of leptons being materialized the reaction is perfectly simple: the lepton and antilepton are produced and move away from each other along the same line (to conserve momentum). It is not so simple for quarks because of the strong force. The strong force will not allow quarks to appear on their own. It is very unlikely that the two quarks will bond to form a meson, as they are produced moving away from each other. They will be linked by a strong force flux tube (see chapter 7) that stretches as they move apart. The quarks will lose kinetic energy as the flux tube gains potential energy. Eventually the flux tube will break up into mesons. The result will be two showers of particles emerging from the reaction point—largely in the direction of the original quarks.

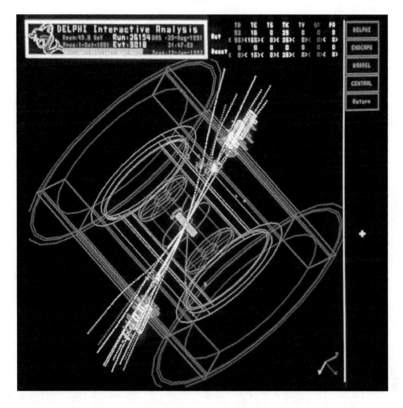

Figure 9.4. A two-jet event seen in the DELPHI detector (see chapter 10). (Courtesy of the CERN press office.)

There will be some divergence; the particles will not appear along two ruler-like lines, but they do retain much of the initial quark's motion. Appropriate equipment can record their paths producing quite striking computer displays such as that shown in figure 9.4.

The streams of particles have been christened *jets* and they are striking visual evidence for the existence of quarks.

There is even more to be learned from the study of jet events.

In chapter 5 we discussed the various threshold energies that are passed as the e^+e^- energy is increased. Each threshold corresponds to there being enough energy to produce a new, more massive, flavour of particle. A threshold is reached at each new quark or lepton mass. This provides a nice way of counting the numbers of quarks that exist.

There is no problem in counting leptons. When they are produced they are not accompanied by jets of particles—they can be seen and studied in isolation. Quarks are a different matter. There is no direct way of telling what flavour of quark is produced by looking at the jets of particles. Their identities are completely swamped by the other hadrons. There is, however, a beautiful way of flavour counting—one simply looks at the relative number of times hadrons are produced as opposed to leptons.

Specifically, physicists study the ratio R:

$$R = \frac{\text{prob}(e^+e^- \rightarrow \text{hadrons})}{\text{prob}(e^+e^- \rightarrow \mu^+\mu^-)}.$$

We can see that the ratio will increase every time a new flavour threshold is passed. All the reactions that were possible below the threshold still continue, but above the threshold there is another reaction possible that will increase the number of events that produce hadrons. Notice that we do not worry about *what* hadrons—that the annihilation produces hadrons *at all* is an indication that a $q\bar{q}$ pair has been formed.

R depends only on the quark charges. Specifically:

$$R = \Sigma(Q_i^2)$$

in which the summation only counts the charges of those quarks whose masses are low enough to be produced at the specific energy.

R is measured at various energies. If the energy is such that only the u and d quarks can be produced, then R should take the value

$$R = (+2/3)^2 + (-1/3)^2 = 5/9.$$

This should then stay constant until the energy is great enough for the s quark to be produced, at which point

$$R = (+2/3)^2 + (-1/3)^2 + (-1/3)^2 = 2/3.$$

Then again, once the c threshold is passed

$$R = (+2/3)^2 + (-1/3)^2 + (-1/3)^2 + (+2/3)^2 = 10/9$$

and so on. Above 10 GeV the u, d, s, c and b quarks are all produced giving an R value of $11/9 = 1.22$. The actual value, as measured by experiments at the PETRA accelerator in Germany, is 3.9 ± 0.3 which is clearly in disagreement with the theory. The solution to this problem lies in another property of quarks which turns out to be the key to understanding the strong force.

9.3.1   Colored quarks

Quarks feel the strong force, whereas the leptons do not. Evidently there is a property of quarks that the leptons do not possess. Baryon number is the obvious candidate, but this does not explain why quarks must bond together as qqq, $\bar{q}\bar{q}\bar{q}$ or $q\bar{q}$ combinations. The correct suggestion was made by Greenberg, Han and Nambu, who realized that there is a strong force 'charge' comparable to the electrical charge that a particle must have to experience the electromagnetic force. Electrical charge comes in two varieties: positive and negative. Their property had to come in three types if it was to explain quark bonding. They named the property *color*[6].

Greenberg, Han and Nambu theorized that quarks can come in three colors: red, blue and green and that any stable hadron must be a colorless combination of quarks. Hence a baryon must contain a red, blue and green quark. A proton, for example, could be u(red)u(blue)d(green), or u(r)u(g)d(b), or u(g)u(b)d(r), etc. Any combination will do as long as there is only one of each. Antiquarks have anticolor: antired, antiblue and antigreen. A meson is therefore a colorless combination of color and anticolor. For example a π^+ would be u(r)$\bar{d}$($\bar{r}$), or a green–antigreen or a blue–antiblue. From this unlikely beginning the whole, successful, theory of the strong force was developed.

Obviously color is not meant in the visual sense. The strong force charge was named color as the pattern of combining three quarks of different charge together is reminiscent of combining primary colours together to make white. The only connection lies in the name. Think of color as being the strong force version of electrical charge.

The relevance of this to our discussion of R is that it multiplies each R value by 3. Whenever the reaction

$$e^+ + e^- \rightarrow q + \bar{q} \tag{9.2}$$

takes place, the resulting quarks could be either red/antired or blue/antiblue or green/antigreen. For each quark flavour there are three distinct quarks that can be produced. The reaction is three times more likely.

If we multiply the 10 GeV prediction by three, then we get $11/3 = 3.67$ which is well within the experimental uncertainty of the measured value. Not only does color form the basis of a theory of the strong force, but it is obviously required to make the theoretical R values match the measured values.

9.4 The November revolution

When the quark hypothesis was first put forward only three quarks, the u, d and s, were required to explain all the hadrons that had been discovered. Hence the list of elementary particles known to physicists comprised u, d and s quarks along with e^-, ν_e, μ^- and ν_μ; three quarks and four leptons. This apparent asymmetry between the number of leptons and quarks led Glashow and Bjorken to suggest, in 1964, that there might be a fourth quark to even up the numbers. They named this hypothetical object the charm quark. At the time there was no experimental evidence to justify such an extension. Later Glashow, Iliopoulos and Maini used the idea to provide an explanation for the non-occurrence of the decay $K^0 \rightarrow \mu^+ + \mu^-$. The explanation is technical and beyond the scope of this book, but it did something to legitimize the suggestion of a fourth quark.

The real revolution came in November 1974 when two teams of experimentalists announced the independent discovery of a new and unexpected type of meson (remember all the other mesons could be accounted for from the u, d and s quarks). One group named the particle the J and the other the ψ. Many people still refer to it as the J/ψ. A ready-made explanation existed for this new particle—it was the lowest mass version of the $c\bar{c}$. Acceptance was not immediate, but it was the only explanation that did not have serious drawbacks. Any lingering doubts about the existence of the charm quark were dispelled during 1975 and 1976 when more particles containing the charm quark were discovered. In 1976 Samuel Ting and Burton Richter (the leaders of the two groups) were awarded the Nobel Prize for their joint discovery.

As one might imagine, the story did not end there.

A new lepton, the tau, was discovered in 1975 (for this discovery Marty Perl received a share of the 1995 Nobel Prize) that once again destroyed the symmetry between the number of quarks and leptons. Physicists immediately went on alert to look for a new quark. On 30 June 1977 Leon Lederman announced the discovery at Fermilab of the upsilon meson that turned out to be the $b\bar{b}$.

Evidence for the tau-neutrino is indirect but convincing and if nature requires equal numbers of quarks and leptons then the top quark must exist. In 1995 Fermilab announced that a two-year run of experiments produced evidence for the top quark. Top had been found after an eighteen-year wait.

The argument for equal numbers of quarks and leptons is not simply that of elegance ('if I were God, that is how I would do it!') but is also required by some theories that go beyond the standard model. However, the reader is entitled to ask how we can be sure that the progression stops at the third generation. Recent evidence produced at CERN has enabled us to 'count' the number of generations and establish that there are only three. We shall refer to this again in chapter 11.

9.5 Summary of chapter 9

- The existence of quarks was originally a theoretical idea that brought order to the pattern of hadron properties;
- the deep inelastic scattering experiments can be interpreted as evidence for high speed electrons interacting with localized charges within protons and neutrons;
- jets can be explained by the materialization of quarks and antiquarks out of the electromagnetic energy produced by e^+e^- annihilations;
- the R parameter is a means of counting the number of quark flavours that can be produced in e^+e^- annihilations as the energy of the reaction is increased;
- the R parameter provides evidence for the existence of a new quark property known as *color*;
- color plays the role of 'charge' in the strong force that electrical charge plays in the electromagnetic force;

- the discovery of the J/ψ meson was evidence for the charm quark that had been suggested in order to balance the numbers of quarks and leptons;
- more recent discoveries have taken us to the current state of six confirmed leptons and six quarks;
- there are some sound theoretical reasons for believing that there must be equal numbers of quarks and leptons.

Notes

[1] Gell-Mann named them quarks, Zweig coined the term aces. Quarks won, perhaps because there are four aces in a pack of cards—in which case the name was slightly prophetic.

[2] Later to become famous as the inventor of the Geiger counter.

[3] In 1914 Rutherford described it, perhaps uncharitably, as being 'not worth a damn'.

[4] The discovery of quantum mechanics, the physics of motion that must be applied within atoms, has shown that electron 'orbits' are far more complicated than planetary orbits.

[5] The effect is increased by the relativistic time dilation. A fast moving object will always see other particles to be moving more slowly as external time slows down when you are moving close to the speed of light.

[6] I will follow the American spelling to emphasize that this is *not* the colour that we see with our eyes.

Chapter 10

Experimental techniques

In this chapter we will look at various devices that are used to accelerate and detect particles. Modern accelerators are reaching the practicable limit of what can be achieved with the techniques that are in use. It is becoming clear that if we wish to work at energies very much higher than those that are currently available, then some new technique for accelerating the particles must be found.

10.1 Basic ideas

When a particle physics experiment is performed the procedure can be split into three stages:

1. *Preparing the interacting particles*
   This may involve producing the particles, if they are not common objects like protons or electrons; accelerating them to the right energies and steering them in the right direction.

2. *Forcing the particles to interact*
   This involves bringing many particles within range of each other so that interactions can take place. This must happen frequently so that large amounts of time are not wasted waiting for interesting interactions to take place.

3. *Detecting and measuring the products*
   The particles produced by the interaction must be identified and

their energies and momenta measured if the interaction (which is not directly observed) is to be reconstructed.

These three stages are, to a degree, independent of each other. Each poses its problems and has forced technology to be improved in specific ways so that modern particle physics experiments can work.

10.2 Accelerators

10.2.1 Lawrence's cyclotron

The first real particle accelerator was developed by Ernest Orlando Lawrence, an associate professor at Berkeley, between 1928 and 1931. His *cyclotron* was the first device to combine the use of an electrical field to accelerate the charged particles with a magnetic field to bend their paths.

Figure 10.1 shows a diagram of the first cyclotron built to Lawrence's design. The device consists of two hollow 'D' shaped pieces of metal placed between the circular poles of an electromagnet. A small radioactive source placed at the centre of the device emits charged particles. The paths of these particles are deflected by the magnetic field (see page 22) and they move into circular arcs passing from inside one D into the other. While inside the volume of a D the charges experience no electrical force, even though the metal is connected to a high voltage. However, as the particles cross the gap between them they move from the voltage of one D to the different voltage of the other. This change in voltage accelerates the particles as they cross the gap.

To make sure that the particles are accelerated rather than slowed down by the change in voltage, the D that they are moving into has to be negative with respect to the D they are emerging from (if the particle is positive).

The particle now moves along a circular path through the second D, but at a slightly higher speed. When it crosses back into the first D at the other side of the circle, it experiences a change in voltage again. However, this time the voltage would be the wrong way round to accelerate the particle and it would be slowed down. The trick is to arrange to swap the sense of the voltage round while the particle is inside the second D. Consequently, by the time it arrives at the gap, the voltage

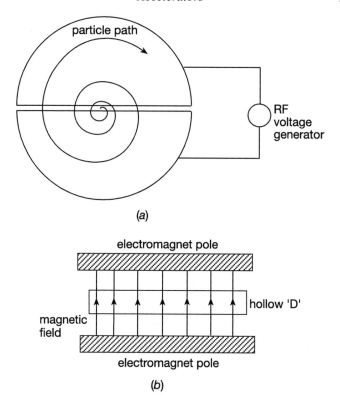

Figure 10.1. Lawrence's cyclotron: (*a*) plan view, (*b*) cross section.

is always in the right direction to accelerate the particle and it receives a small kick every time it crosses a gap (twice per revolution).

Organizing the voltage to swap over at the right time would be difficult if it were not for a curious property of the particle's motion. Every time the particle is accelerated, you would think that it arrived at the gap a little faster than last time. However, as the speed of the particle increases so does the radius of the circular path. This means that a faster particle is deflected into a bigger circle and so has further to travel. Even though it is moving faster, it still arrives at the gap at the same time. Hence the changing voltage can be switched at a constant rate and it will always match the motion of the particle.

length of path $= 2\pi r$

where r is the radius of the circle; the time taken is

$$T = \frac{2\pi r}{v}$$

however, $$r = \frac{mv}{Bq}$$ (page 24)

$$\therefore \quad T = \frac{2\pi mv}{Bqv}$$

$$= \frac{2\pi m}{Bq}$$

so T is independent of the speed at which the particle is moving. The frequency with which the field should be switched over is

$$f = \frac{1}{T}$$

$$= \frac{Bq}{2\pi m}$$

f has to be at radio frequencies (RF) for a typical subatomic particle.

The first cyclotron measured 13 cm in diameter and accelerated protons to 80 keV energy. The device could be made bigger to achieve more energy, but eventually two fundamental limitations were met. Firstly, the circular magnets were becoming impracticably large, and secondly the particles were being accelerated to relativistic speeds and the constant time to complete a revolution no longer applied. This places a limit of about 30 MeV on a proton cyclotron.

The first development was to alter the frequency of the RF field to keep in step with the accelerating particles. This means that the accelerator can no longer work with a continuous stream of particles. Particles at a later stage of acceleration would be moving faster than those that had just entered the accelerator, and so it would be impossible to keep the field in step with all of them at the same time.

The answer was to accelerate particles in bunches to the edge of the machine, then extract them before a new bunch entered the accelerator.

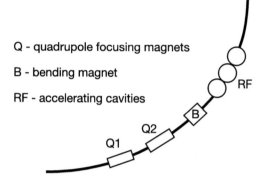

Q - quadrupole focusing magnets

B - bending magnet

RF - accelerating cavities

Figure 10.2. Part of a modern synchrotron.

While the bunch was being accelerated the radio frequency (RF) was adjusted to synchronize with them crossing the accelerating gap. The device was called a *synchrocyclotron*. Modern accelerators still accelerate particles in bunches. In 1946 a synchrocyclotron accelerated deuterium nuclei to 195 MeV.

Although the synchrocyclotron overcame the problem of the relativistic limit, it did not address the issue of the magnets.

As the demand for higher energies continued it became harder to increase the strength and size of the magnets. The next step was to use a series of magnets round an accelerating ring rather than two large magnetic poles.

In this design of machine the particles are kept on the *same* circular path by increasing the strength of the magnetic field as they become faster. The accelerating RF fields are applied at several points along the ring and the frequencies varied to keep in step with the particles. Such a machine is called a *synchrotron* and is the basic design followed by modern particle accelerators throughout the world.

10.2.2 A modern synchrotron

Figure 10.2 shows the most advanced form of particle accelerator now in use. A variety of different magnets bend and focus the beam with radio frequency cavities to accelerate it.

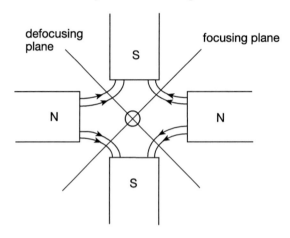

Figure 10.3. A quadrupole magnet. The beam runs perpendicular to the page through the circle at the centre of the diagram.

Quadrupole magnets

The quadrupole magnets (figure 10.3) are designed to kick particles that are drifting sideways out of the bunch back into line. They can only do this in one plane and they tend to move particles out of line slightly in the perpendicular plane. To overcome this they are used in vertical and horizontal pairs.

Bending magnets

Bending magnets are often made with a superconducting coil to allow large magnetic field to be maintained without expensive electricity bills. These magnets have their fields steadily increased as a bunch is accelerated to keep the particles in the same circular path.

RF cavities

The RF cavities are designed to accelerate the particles as they pass from one cavity into the next. They are designed to switch their potentials at radio frequencies keeping the bunch of particles accelerated as they pass through the system. One of the largest accelerating rings in the world was the Large Electron Positron Collider (LEP) at CERN in Geneva. LEP used 1312 focusing magnets and 3304 bending magnets round its 27 km circumference and accelerated e^+ and e^- to an energy of 50 GeV each. The e^- and e^+ were sent in opposite directions round the same ring to collide at several points where detectors measured the reaction products.

Although the basic synchrotron ring structure dominates current particle accelerator design it does have some disadvantages.

Disadvantages of synchrotrons

- Bending magnets are reaching the limits of their field strength, so to go to higher energies very large rings will have to be constructed. LEP's 27 km is nearing the practical limit for such a ring—in its construction new surveying techniques had to be developed. Superconducting magnets are capable of much higher field strengths. They require a great deal of 'plumbing' to supply them with the coolant necessary to keep the magnets at superconducting temperatures. Again there is a practical limit to the size of the ring that can be supplied with superconducting magnets.
- All circular accelerators suffer from *synchrotron radiation*. Any charged particle moving in a circular path will radiate electromagnetic radiation. The intensity of this radiation depends inversely on the fourth power of the particle's mass ($\sim 1/m^4$) and inversely on the radius of the path ($\sim 1/r$). Consequently electron rings produce much more radiation than proton rings and large diameters are to be preferred. The energy lost by particles in producing this radiation has to be replaced by accelerating the particles again[1].

10.2.3 Linacs

A linear accelerator (linac) in which the particles move along a straight line through a series of accelerating cavities, does not suffer from synchrotron radiation. This makes them very useful in providing a preliminary acceleration for particles before they are inserted into a big ring.

However, there is one existing linear accelerator that is in use in its own right. The Stanford Linear Accelerator Centre (SLAC) uses a 3 km long linear accelerator that accelerates both e^- and e^+ to 50 GeV along the same line and then bends them round at the end to collide with each other. In its previous guise the SLAC linac was used to accelerate e^- to 20 GeV for use in the deep inelastic scattering experiments.

The obvious disadvantage of linacs (their length if they are to achieve high energies) places a severe limit on how much more they could be developed.

10.2.4   List of current accelerators

- *CERN: European Particle Physics Laboratory, Geneva*
  (i) LEP: collided e^+ with e^- at 100 GeV upgraded to 180 GeV, decommissioned in November 2000 after 11 years of use, 27 km tunnel
  (ii) LHC: under construction, will collide p and $\bar{p}$ at 14 000 GeV using the LEP tunnel
- *FERMILAB: Chicago*
  Tevatron: collides p and $\bar{p}$ at 1800 GeV
- *HERA: Hamburg, Germany*
  collides 820 GeV p with 26.7 GeV e^-
- *SLAC*
  (i) SLC: collides e^+ with e^- at 100 GeV at the end of the 3 km linac
  (ii) PEP: e^+e^- ring colliding at 30 GeV—discontinued now; PEP II was to be commissioned in 1999 and tuned to produce b quarks.

10.3   Targets

There are two ways of forcing particles to interact: by using a fixed target or colliding them with each other. Each has its own specific advantages and disadvantages.

10.3.1   Fixed targets

In a fixed target experiment, one of the interacting particles is contained in a block of material which acts as a target into which a beam composed of the other particles is fired. The target particle is assumed to be at rest while the accelerated energy and momentum are in the beam particle. Targets can be blocks of metal, tanks of liquid hydrogen that form a proton target (called bubble chambers) or silicon vertex detectors (see later in section 10.4). The latter two fixed targets have the advantage of being able to reproduce particle tracks as well as acting as providing particles for the reaction.

Advantages

- Easy to produce many interactions as the beam particles have a high probability of hitting something as they pass through the target;
- by using a bubble chamber or vertex detector it is possible to see the point in the target at which the interaction took place;
- fixed targets provide a convenient way of producing rare particles that can themselves be formed into a beam (e.g. kaons, antiprotons, pions etc).

Disadvantages

- As the beam particle carries momentum and the target particle has none, there is a net amount of momentum that must be conserved in the reaction. This means that the produced particles must carry this momentum. If they have momentum, then they must also have kinetic energy. So not all the energy in the reaction can be used to create new particles. This sort of experiment is not an efficient use of the reaction energy.

10.3.2 Colliders

In this type of experiment, two beams of accelerated particles are steered towards one another and made to cross at a specified point. The beams pass through each other and the particles interact.

Advantages

- It is easier to achieve higher energies as both particles can be accelerated rather than having to rely on just the one to carry all the energy;
- if the beams collide head on and the particles carry the same momentum, then the total momentum in the interaction is zero. This means that all the produced particles must have a total of zero momentum as well. This is a much more efficient use of energy as it is possible for all the reaction energy to go into the mass of new

particles (if they are produced at rest)—no kinetic energy is needed after the reaction.

Disadvantages

- It is difficult to recreate the actual point at which the particles interacted—which is necessary if momentum and direction are to be measured and short lived particles (which may not travel very far from that point before they decay) are to be seen;
- it is difficult to ensure that the particles hit each other often enough to produce an adequate supply of interactions. This is measured by the *luminosity* of the experiment—the number of reactions per second per cm^2 of the beam areas. Modern experiments have achieved improved luminosity over the earliest colliders (the greatest luminosity achieved so far has been at Fermilab: 1.7×10^{31} s^{-1} cm^{-2}). This improvement and the more efficient use of energy over fixed target experiments makes the collider the experiment of choice in modern particle physics.

10.4　Detectors

All particle detectors rely on the process of ionization. When a charged particle passes through matter it will tear electrons away from the atoms that it passes. This will result in a free electron and a positive ion in the material. The number of ions formed by the particle is a measure of its ionizing power and this, in turn, depends on the charge and velocity of the ionizing particle. Clearly a neutral particle has an extremely small chance of ionizing atoms and is consequently very difficult to detect.

The way in which the ions formed are used to track the progress of particles depends on the detector that is being used. Although there are many detectors that have been specifically designed to operate in a certain experiment, there are some general purpose devices that are in common usage.

10.4.1　The bubble chamber

These devices have gone out of fashion now partly as they are not suitable for use in a collider experiment, but also because they cannot cope with the number of reactions per second required in modern experiments, which tend to deal with very rare reactions.

However, bubble chambers used to be extremely important. Their considerable advantage was that they acted as a target as well as a particle tracker.

A bubble chamber consists of a tank of liquid hydrogen that has been cooled to a temperature at which the hydrogen would normally boil. The liquid is prevented from boiling as the tank is placed under pressure. In this state the beam of particles is allowed into the tank and after a short time (to allow an interaction to take place) the pressure is released. Charged particles that have crossed the chamber will have left a trail of ionization behind them. As the pressure has been released local boiling will take place within the tank. The bubbles of gas will tend to form along the ion trail as the charged ions act as nucleation centres about which the bubbles form. These bubbles are allowed to grow for a certain time, and then a photograph is taken of the chamber that provides a record of the interactions. There is a compromise in how long the bubbles are allowed to grow before the photograph is taken. A short time means small bubbles, which are hard to see, but which allow precision measurements to be taken. A long time gives larger, clearer bubbles but the extra space that they take up means that it can be difficult to separate tracks that run close to each other. A typical bubble size would be a few microns.

Advantages

- The actual interaction point is visible;
- if the chamber is placed in a magnetic field the momentum of the particles can be measured by the bending of the tracks;
- the thickness of the tracks is related to the amount of ionization which can be an indication of the type of particle that made the tracks;
- the existence of neutral particles can be deduced if the particle decayed into charged particles within the chamber (see figure 10.4)
- can be very accurate (a few microns on a track).

Disadvantages

- The chamber has to be re-set by compressing the liquid which bursts the bubbles. This means that there is a 'down time' during which

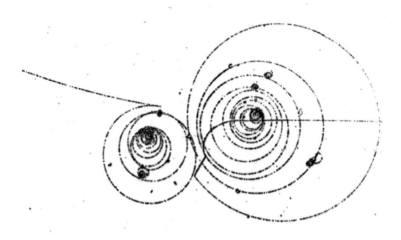

Figure 10.4. A typical bubble chamber picture. (Courtesy of the CERN press office.)

any interactions that take place are not recorded. A very fast bubble chamber can go through 30 expansion cycles in a second, but this is nowhere near the rate at which events can be recorded with modern electronic detectors;

- does not produce instant information about tracks;
- needs a great deal of extra equipment to control the temperature, take the photographs and regulate the pressure.

10.4.2 Scintillation counter

A charged particle passing through a scintillating material will cause the material to emit a pulse of light. This light is produced by ionization of electrons which then recombine with an ion, excitation of electrons within the atom or even the breaking up of molecules (depending on the material). The light can be collected and detected using a sensitive photocell. This sort of chamber is used simply to detect the presence of particles and gives no information on the direction of travel.

10.4.3 Čerenkov detector

A Čerenkov detector is a more advanced form of scintillation counter. A particle passing through a material at a velocity greater than that at which light can travel through the material[2] emits light. This is similar to the production of a sonic boom when an aeroplane is travelling through the air faster than sound waves can move through the air. This light is emitted in a cone about the direction in which the particle is moving. The angle of the cone is a direct measure of the particle's velocity. A system of mirrors can ensure that only light emitted at a given angle can reach a detector, so that the Čerenkov chamber becomes a device for selecting certain velocities of particle. Alternatively, if the momentum of the particle is known (from magnetic bending) the Čerenkov information on the particle's velocity enables the mass to be deduced so that the particle can be identified.

10.4.4 The multiwire proportional chamber (MWPC)

This detector has a precision nearly as good as that of a bubble chamber, but is able to work at a much faster rate as the information is recorded electronically.

A typical MWPC consists of three sets of wires in parallel planes close to each other (see figure 10.5). The outer two planes are connected to a negative voltage of typically -4 kV and the central plane is connected to earth (0 V). The whole collection is housed in a container that is filled with a low pressure gas.

The sense wires that form the central plane are spaced at 2 mm intervals and of a very small diameter (typically 20 μm). This has the effect of producing a very concentrated electrical field near the wire (figure 10.6).

A charged particle passing through the MWPC will leave a trail of ionized gas particles behind. Under the influence of the electrical field the electrons from these ions will start to drift towards the nearest wire. As they get close to the wire they are rapidly accelerated by the strong field. The electrons are now moving fast enough to create ions when they collide with gas atoms. The electrons from these ions will also be accelerated and the process continues. Consequently a few initial ions can rapidly produce a substantial charge at one of the wires producing a

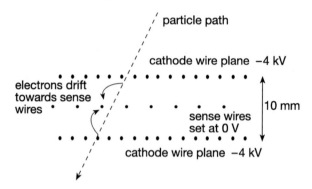

Figure 10.5. A typical MWPC cell.

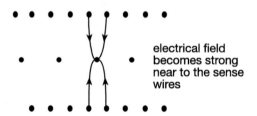

Figure 10.6. The field near a sense wire.

measurable current pulse along the wire. This is known as the *avalanche effect*.

Electronic amplifiers connected to the sense wires can easily detect which wire produces the most current and so say which wire the particle passed closest to. They are also capable of providing information on more than one particle passing through the chamber at once.

An MWPC working in this fashion can produce a signal within 10^{-8} seconds of the particle passing through. By using a series of such detectors with the wire planes at angles to one another it is possible to reconstruct the path of the particle in three dimensions.

MWPCs have revolutionized particle physics experiments with their fast detecting times. Modern experiments produce huge numbers of interactions—far too many to be recorded. For this reason computers

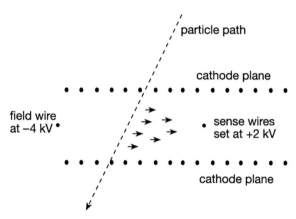

Figure 10.7. The drift chamber.

have to have rapid information about the nature of the interaction. MWPCs can pass information on to a computer within microseconds. The computers can then carry out a preliminary analysis of the event (typically taking a few microseconds) to decide if it is worth recording.

10.4.5 Drift chamber

Drift chambers measure the time taken for ions to arrive at the sense wires. A typical design of drift chamber is shown in figure 10.7. The central wire plane is set up with alternate field and sense wires and the voltages carefully chosen to produce a uniform drift speed for electrons along the chamber. This is a very delicate operation. The field strength between the wires has to be adjusted to compensate for the electron's loss of energy due to collisions with gas atoms within the chamber. If this is done properly the electrons will drift at a constant rate to the sense wire at which they will trigger an avalanche producing a detectable current pulse. As the drift speed is known, the drift time can be used to calculate the distance that the electrons had to move and so the distance of the particle's path from the sense wire is measured. With this sort of technique measurements are possible with an accuracy of a few tens of microns.

If the drift chamber is to work correctly another device must be used to start the clock running on the drift time when a particle enters the chamber. Typically this is a variety of scintillation counter.

10.4.6 Silicon detectors

The pressure of modern experimentation with very large numbers of events per second to be analysed and tracks recorded has led to the need for very high accuracy tracking equipment that can be read electronically.

Silicon detectors consist of strips of silicon (which is a semiconducting material as used in integrated circuits) which are very narrow (tens of microns) running parallel to each other. When a charged particle passes through the silicon it causes a current pulse in the strip that can be detected with suitable electronics.

The most up to date versions of these detectors are 'pixelated' which means that they are broken up into very small dots of silicon the way the photodetector in a camcorder is. Detecting which strip or pixel recorded a 'hit' and tracking these hits through a sequence of silicon detectors can give a very precise reconstruction of a particle's track. Micron accuracy is easily achieved. In addition, as these detectors are very fast and can be read directly by a computer they can provide information about an event that is virtually immediate, enabling the software that is controlling the experiment to decide if the event is worth permanently recording (on magnetic tape or optical disk) before the next event takes place.

10.4.7 Calorimeters

These are devices for recording the amount of energy carried by a charged particle. There are two types of calorimeter which work in different, but related, ways.

Electromagnetic calorimeters

These are designed to measure the energy carried by electrons, positrons or high energy photons. There are many different ways of constructing a calorimeter. One design consists of sandwiches of lead and a plastic scintillator. When an electron or positron passes through the lead it is frequently deflected by the atoms in the material. This can lead to the emission of high energy photons as the electron or positron loses energy in the process. The amount of lead is chosen to ensure that virtually all electrons or positrons in the experiment are brought to rest by radiating away all their energy in this manner. The high energy photons interact with the atoms of lead and this generally leads to their conversion into an electron–positron pair. These will, in turn, radiate more photons as they are brought to rest in the lead. In this way, an initial electron or positron can create a shower of such particles through the calorimeter. As this shower passes through the scintillator it causes flashes of light that can be detected by sensitive electronic equipment. How far the shower penetrates through the sequence of lead sheets and scintillator is directly related to the energy of the initial particle, as is the total number of particles produced in the shower. Photon energy can be measured as well, as high energy photons entering the lead will convert into electron–positron pairs and this will also set up a shower. The ability to detect and measure photon energy is often important in trying to pick up π^0's which decay into a pair of photons.

Hadron calorimeters

These are designed to measure the energy carried by hadrons. They also work by generating a shower from the original particle, but the process is not electromagnetic. One design uses iron sheets instead of the lead and the hadron reacts with an iron atom to produce more hadrons, which in turn react with other atoms. A shower of hadrons passes through the calorimeter setting off signals in the scintillator as they pass. The distance that the shower penetrates into the calorimeter gives the energy of the initial hadron.

10.5   A case study—DELPHI

The DELPHI detector was one of the four detectors specially designed and built for experiments at the LEP accelerator in CERN. DELPHI ran from 1991. The first stage of the LEP accelerator collided electrons and positrons at 45 GeV per beam and so the total energy in the collision was exactly that required to produce a Z^0. Later the beam energy was increased (LEP2) to be enough to produce W^+ and W^- particles:

$$e^+ + e^- \rightarrow Z^0 \qquad \text{LEP1}$$
$$e^+ + e^- \rightarrow W^+ + W^- \qquad \text{LEP2.}$$

The purpose of LEP was to provide a tool that could probe our understanding of the standard model to very high precision. Between

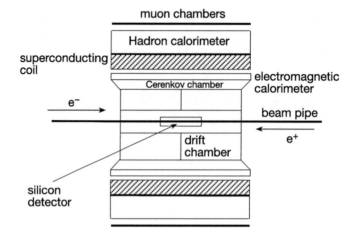

Figure 10.8. Part of the DELPHI detector.

1991 and 1994 LEP produced something like 4 million Z^0 events in the DELPHI detector.

Figure 10.8 shows the central part of the DELPHI detector. Other components fitted on to either end to provide a complete coverage of the region round the collision point, which was at the centre of the detector. The whole detector, when assembled, was 10 m long and had a radius of 5 m.

The silicon detector is too small compared to the rest of the equipment to be shown on the diagram. It had a radius of 5.2 cm and was designed to locate the interaction point and to feed information to a computer allowing the tracks coming from the interaction to be reconstructed.

The large drift chamber (3 m × 1.2 m radius) was divided into two halves vertically through the collision region. Ions produced by particles passing through the chamber drifted towards the ends farthest from the interaction plane. The chamber provided accurate tracking data and, by measuring the amount of charge produced, some information helping to identify the particles passing through it. Particle identification was also provided by a complex Čerenkov chamber which was designed to help distinguish between protons, pions and kaons.

Figure 10.9. The drift chamber being moved into place in DELPHI. Note the layers of calorimeter in rings round the drift chamber. (Courtesy of the CERN press office.)

The electromagnetic calorimeter consisted of layers of lead wires with gaps between them. An electromagnetic shower built up as the electrons and positrons travelled out from the centre of the detector through each layer of lead wires. In the gap between layers the shower ionized a gas. The ions then drifted along a layer under the influence of an electric field to sense wires at the end of the layer. Detecting the amount of charge arriving at the end of each layer charted the progress of the shower and so the energy of the initial particle.

Figure 10.10. The DELPHI experimental hall. (Courtesy of the CERN press office.)

The complete calorimeter was divided into rings along the length of the detector, each ring having 24 separate modules. This was to ensure that the chances of more than one shower being developed in a module at the same time were minimized.

The whole central part of the detector was in a magnetic field produced by the superconducting coil. This was the largest superconducting magnet in the world: 7.4 m long and 6.2 m in diameter. The field ran parallel to the beam pipe and the coil carried a current of 5000 A. To maintain its low resistance the coil had to be kept at a temperature of 4.5 K.

The final parts of the detector were the hadron calorimeter and the muon chambers. The iron in the hadron calorimeter was designed to stop all the hadrons produced in the interaction, so that anything passing into the muon chambers was bound to be a muon—the only particle that could get through the rest of the detector.

10.6 Summary of chapter 10

- The modern particle accelerator is a synchrotron;
- quadrupole magnets are used to focus the bunch of particles that are accelerated;
- bending magnets keep the particles in a circular path;
- radio frequency cavities accelerate the particles;
- circular ring accelerators suffer from synchrotron radiation (especially electron accelerators) requiring them to have a large diameter;
- linear accelerators do not produce synchrotron radiation, but have to be very long to accelerate particles to high energies;
- fixed target experiments produce lots of interactions, but are wasteful of energy due to the need to conserve momentum;
- collider experiments can be designed to have zero net momentum and so are much more energy efficient, but they are harder to produce lots of interactions with;
- there are a variety of detectors for identification (Čerenkov, drift chamber) and measurement (bubble chamber, MWPCs, drift chambers);
- calorimeters are designed to measure a particle's energy by the number of shower particles that it can produce when passing through dense matter.

Notes

[1] Even with a modest electron ring like that at DESY in Germany, which accelerates electrons to 20 GeV and has a radius of 256 m, 8 MW of electrical power are required to replace the energy lost due to synchrotron radiation alone.

[2] Before I get letters from the people who think that this violates relativity, I should explain that the particle must be moving faster than a light wave could if the light wave were to pass through the material. Light passing through matter

never moves in an exactly straight line as it is scattered by the atoms. This means that a beam of light crossing a block of glass has to travel much further than the width of the block to reach the other side. We interpret this as the light moving more slowly in the block—hence refractive index. A charged particle moving through the material can get to the other side more quickly—hence it appears to move faster than light. The relativistic limit on velocity is the speed of light in a vacuum.

Interlude 1

CERN

CERN (the Conseil Européen pour la Recherche Nucléaire) is the largest research facility of its type in Europe and one of the largest in the world. The idea of establishing a major research facility that could be used by all European countries arose shortly after the end of the second world war. The aim was partly political (it was seen as a step towards reconciliation by working together) and partly to halt the drain of scientists away from Europe. There was also a recognition that advanced experimental science was going to be increasingly expensive. No single country could afford to fund a worthwhile facility, but by combining their efforts financially and intellectually, Europe could play a significant role in fundamental research.

In June 1950 UNESCO adopted CERN as a central part of its scientific policy, the preparatory work having been done under the auspices of the European Centre for Culture. CERN was born on 1 July 1957 when the CERN convention was signed in Paris. In 1955 a 42 hectare site near the village of Meyrin close to Geneva was given over to the new facility. The site was extended by 39 hectares in 1965 by agreement with France and from that time on CERN has straddled the Swiss–French border. Another extension on French soil took place in 1973 when a further 90 hectares were made available.

CERN's main function is to provide accelerated beams of particles that can be used for experimental purposes. It develops and maintains particle accelerators and research facilities as well as the support staff required. Collaborations of physicists from various universities apply

Figure I1.1. A view of CERN showing the LEP ring. (Courtesy CERN.)

to run experiments on the site. CERN itself is funded by direct grant from the various member states. Money for the experiments comes from the universities taking part via their government research grants. Experiments can take many forms, from the small and cheap experiment that manufactured anti-atoms of antihydrogen (PS210—see interlude 2) to the very high cost and complex LEP detectors such as DELPHI. In the latter case an experiment can take more than 10 years from initial design through to construction and data-taking followed by the careful analysis of the results. Such projects are a considerable investment in manpower as well as finance. The detectors are designed to be as flexible as possible. Many pieces of research are going on at once using the same equipment.

The careful development of facilities has been one of the major reasons why CERN has been so successful. As our understanding has improved

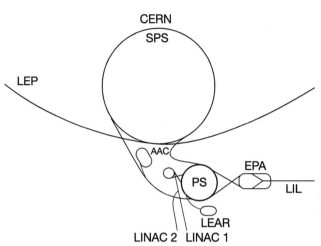

Figure I1.2. The accelerator complex at CERN.

so the need to produce particles at higher and higher energies has driven the development. CERN has been able to use existing accelerators as part of the expansion. Nothing has been 'thrown away' if it could be made to work as part of a new machine. Figure I1.2 shows the current machines that are at work in CERN.

- The PS, or Proton Synchrotron, was the first large accelerator to be built at CERN. It is 200 m in diameter and first started operation in 1959. The PS is able to accelerate particles up to 28 GeV. Most recently it has been used as a pre-accelerator for LEP and the SPS. It is fed with protons or ions made from heavy nuclei via two linear accelerators and a 50 m booster ring that takes their energy up to 1 GeV. After acceleration the protons can be passed on to the SPS. Alternatively the PS can take antiprotons from the antiproton accumulator (AAC) and inject them into the SPS.
- The SPS, or Super Proton Synchrotron, was commissioned in June 1976 to accelerate protons up to 400 GeV. It was later upgraded to 450 GeV. The ring is 2.2 km in diameter and runs underground at a depth of between 23 and 65 m. Between 1978 and 1981 the SPS was converted to enable it to accelerate protons and antiprotons simultaneously in opposite directions. These particles were then collided in the experimental halls. While working in this manner

the SPS collider enabled the discovery of the W and Z particles in 1983.

- The Antiproton Accumulator (AAC) was designed to supply the SPS with antiprotons in its collider mode. Antiprotons have to be made in particle collisions and then stored until enough of them can be put into a bunch to be accelerated by the SPS. The AAC stores the antiprotons by circulating them in a 50 m ring.
- LIL is the linear injector for LEP. It is 100 m long and accelerates both electrons and positrons up to 0.6 GeV. The particles build up in the EPA (electron–positron accumulator) and are then sent to the PS that accelerates them to 3.5 GeV. They are then taken up to 22 GeV by the SPS from which they are injected into LEP.
- The LEP tunnel is 27 km in circumference and runs underground at a depth between 50 and 150 m. The LEP accelerator started operation in July 1989 and was finally turned off in November 2000. Initially it collided electrons and positrons at an energy designed to produce large numbers of Z^0 particles. In a subsequent upgrade to LEP2 the energy was boosted to allow W^+ W^- pairs to be produced.
- LEAR (low energy antiproton ring) was designed to take antiprotons from the AAC and decelerates them to 0.1 GeV. This has enabled several unique experiments to be carried out. It was used to supply antiprotons for an experiment to make anti-atoms of antihydrogen (see interlude 2).

LEP and the future

When I started writing the first edition of this book the LEP ring was about to be switched on. After some years LEP turned into LEP2 and in November 2000 the whole thing was shut down.

LEP has been an extraordinary success in many ways. That nearly a quarter of the world's particle physicists should work together on designing, building and using an accelerator and its detectors is a triumph of international collaboration on a scale that puts the UN to shame.

On of the main design aims behind LEP was to test the standard model. In this regard it has been most successful, as the standard model has survived in remarkably good order. In addition persuasive evidence has

Figure I1.3. A view of the LEP tunnel. The beam travels in a pipe inside the multiple magnets that run down the centre of the picture. To the left is the track for a monorail that runs the length of the tunnel. To the right are various servicing and cooling ducts. (Courtesy CERN.)

been gathered that there are only three generations of quarks and leptons. Most recently and dramatically LEP2 started to see some evidence for the Higgs particle (a long sought after object discussed in chapter 11) just as it was about to be switched off.

LEP is currently being replaced by CERN's latest accelerator project–the LHC (Large Hadron Collider). This accelerator will run in the same tunnel as the LEP ring does now (saving on construction costs). Even so, the European collaboration has had to fight hard to gain approval for construction.

The intention was that LHC would become operational in 2004 to collide protons at 9.3 TeV. Later, in 2008, a second stage of construction would enable the machine to reach 14 TeV. However, at a meeting in December 1996, the CERN council agreed a drastic change to this programme. Now the accelerator is due to be completed in one stage by 2005. This has been made possible by a substantial series of contributions made by non-member countries (i.e. countries that do not contribute to the annual running budget of CERN) negotiated during 1996:

Japan	8.85 billion Yen
India	$12.5 million
Russia	67 million Swiss Francs
Canada	$30 million in kind
USA	$530 million in contributions to LHC experiments and the accelerator.

Moving outside the normal CERN circle to obtain finance was an important decision, but it was made in the shadow of a demand from the member states for a reduction in their annual contributions. It was decided that the LHC funding should continue as planned, but that there should be a rolling programme of reductions in the total budget for the member states:

7.5% in 1997, 8.5% in 1998–2000, and 9.3% in 2001 and thereafter.

The LHC will become an accelerator of world significance. It would be a major shock if the Higgs particle did not turn up at this energy (especially given the possible sighting at LEP). The so-called 'supersymmetric' particles, which are expected to have masses of the order of 1 TeV, should also fall within the LHC's sights.

CERN FACTS

Budget: 939 million Swiss Francs (2001)

Percentage contribution of the member states to the CERN annual budget:

Austria	2.33	Netherlands	4.37
Belgium	2.68	Norway	1.59
Czech Republic	0.53	Poland	1.76
Denmark	1.83	Portugal	1.16
Finland	1.32	Slovak Republic	0.21
France	16.05	Spain	6.93
Germany	22.35	Sweden	2.46
Greece	0.98	Switzerland	3.50
Hungary	0.52	United Kingdom	17.67
Italy	11.76		

Permanent CERN staff numbers (end 2000):
90 research physicists
960 applied physicists and engineers
933 technicians
271 craftsmen
448 office and administration staff

Peak power consumption: 160 MW

Chapter 11

Exchange forces

In this chapter we shall look in more detail at the general theory of forces. This theory has been extremely successful in providing a mathematical understanding of the four fundamental forces and has had a profound influence on the way that we think about the origin of the universe. The price has been the difficulty in extracting from it a clear physical picture of what is happening. At the subatomic level forces are not simply 'pushes and pulls'.

11.1   The modern approach to forces

One of the key insights of modern physics has been the realization of the importance of asking the right question. Unless we are clear about what we wish to achieve with our experiments and theories, nature is so intractably difficult in her responses that we make no progress at all. This was clearly demonstrated in the 1920s and 1930s when physical science ran head-long into the problem of interpreting the results of experiments with atoms. As a result of this problem physicists learnt to change their way of thinking and quantum theory was born. The implications of this theory have still not been fully worked out, but the chief lesson is that we cannot have a fully consistent physical picture of what happens in the subatomic world. We have to make do with stories, comparisons and half-true physical pictures. There have been many books that have tried to translate the mathematics of quantum theory into terms that can be understood by the general reader. We do not have the time to dwell on

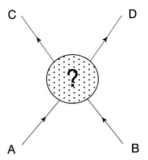

Figure 11.1. The bubble of ignorance.

such matters in this book. However, we will have to face some of the problems as we tackle the nature of forces.

Physicists have taken a realistic look at particle physics experiments and decided upon a theoretical approach that seems very limited in its ambition. However, it does have the virtue of being spectacularly successful!

Figure 11.1 illustrates the fundamental basis of all experiments. Two particles, A and B, enter the experiment having been prepared in a specific way (with a known energy and momentum). They react and as a result two (or more) particles, C and D, emerge into detectors to be identified and have their energy and momentum measured. In between preparation and detection the reaction happens. This is part of the experiment that we cannot look into (symbolized by the bubble on the diagram). We only detect the results of the reaction.

It is impossible to look directly at the reaction, as this would mean interacting with the particles *before* the reaction (in which case they would not be as we had prepared them) or *during* the reaction (in which case, we interfere with the reaction we are trying to study)[1]. For this reason we treat the region in the bubble as a region of ignorance and confine ourselves to trying to calculate the probabilities of different reactions (given a certain A and B, what are the chances of producing C and Ds). Once we have calculated a full set of probabilities for all the possibilities, we can work something out about what might be going on inside the bubble.

Exchange forces

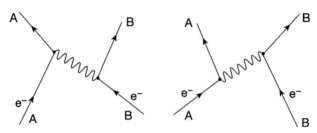

Figure 11.2. One photon exchange.

The first set of reactions to have been fully explored in this manner were those due to the electromagnetic force. The resulting theory is called quantum electrodynamics (QED for short). For the fundamental work on this theory Feynman, Schwinger and Tomonaga were awarded the Nobel Prize in 1965. The theory is highly mathematical (and absolutely beautiful!), but we can gain some insight into its working by considering Feynman's approach to calculations.

11.1.1 Feynman diagrams

QED works by considering all the possible reactions that might take place inside the bubble. It calculates a probability for each possibility and adds them together to give the overall probability. Feynman developed a highly pictorial technique for making sure that all the possibilities were included.

To illustrate this, we will consider a simple reaction:

$$e^+ + e^- \rightarrow e^+ + e^- \tag{11.1}$$

the elastic scattering of electrons. An obvious way in which this reaction may have taken place is illustrated in figure 11.2.

The black blobs on the diagram at the points where the photon is emitted and absorbed are called *vertices*. In figure 1.1 I illustrated such vertices as large black blobs, although in that case it was the weak force that was being discussed.

A partial physical picture of what is happening in these diagrams can be obtained by realizing that, as the electrons approached each other, a

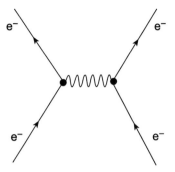

Figure 11.3. A Feynman diagram.

disturbance was set up in their mutual electromagnetic field. The field is not shown on the diagrams, but the disturbance is represented as a photon linking the two particles. One diagram shows particle A emitting the photon that is then absorbed by B. The other diagram has the sequence the other way round. Our discussion of deep inelastic scattering suggests that the photon is formed due to the combined motion of *both* particles— so both diagrams represent what is happening. As there is no way we could experimentally tell them apart, *we have to count both.* For that reason we will abandon all attempts to tell the order in which things happen when they cannot be distinguished and draw a single diagram such as that in figure 11.3.

This is an example of a *Feynman diagram*—there is no time sense in this diagram; it is simply a sketch of a *process that includes both the time orderings in figure 11.2.* Feynman discovered a way of including both time sequences automatically in the same calculation.

This is not the only possible way in which the electrons might scatter off each other. Figure 11.4 shows some more Feynman diagrams that lead to the same result.

Each diagram represents a possibility that must be considered. Feynman derived precise mathematical rules that allow physicists to calculate the probability of each process directly from the diagrams. It is then a matter of adding the results in the correct manner to calculate the overall

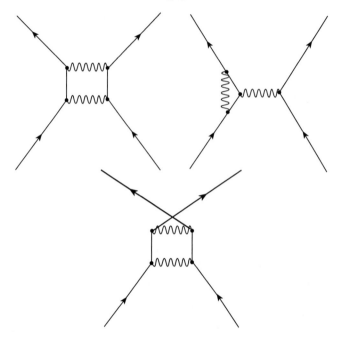

Figure 11.4. Some more complicated Feynman diagrams.

probability of the reaction. Each diagram represents a possibility of what might be happening inside the bubble.

Obviously there are a vast number of Feynman diagrams that can be drawn and would have to be included—apparently making it an impossible task to calculate the final result. Fortunately this is not the case as QED is a 'well behaved' theory. As the diagrams increase in complexity they also decrease in probability. To obtain a result accurate enough to compare with experiment only the first few diagrams need to be included. QED has been checked to 15 decimal places of accuracy on one of its predictions making it the most accurate theory that we have.

It is very natural to ask which of these multiple diagrams represents the truth—surely they cannot all be happening at the same time?

This is the point at which we have to surrender our wish for a physical picture. We know that calculating the probabilities in this fashion produces very accurate results, but we have no way of telling which diagram actually happened. Furthermore, the probability of any one diagram is far less than that which is measured. So it is not a case of one of the diagrams being 'right' in different cases—they are all right every time the reaction happens! All we can say is that the disturbance in the electromagnetic field is very complicated and that each diagram represents an approximation to the actual physical process. Only by adding together all the diagrams can we get a mathematically correct answer[2].

Some books only consider the first diagram in the sequence and develop the idea that the repulsion between the electrons is due to their exchanging a photon—as the diagram seems to suggest. Imagine two people cycling along next to each other. One throws a football towards the other. As a result he/she recoils and changes direction. The football then strikes the other cyclist causing him/her to be knocked away from his/her initial path. The photon is supposed to do a similar job to that done by the football.

While this is a quaint and easily visualized model, it does lead to severe problems if it is followed too far. For example, it becomes very hard to explain attraction.

11.1.2 The problem of attraction

Many people rapidly work out that the exchange football idea is inadequate to explain a force *pulling* particles together. However, when you draw the first Feynman diagram for the interaction between an e^+ and an e^- it looks exactly the same as figure 11.3 suggesting that the same process is at work.

Figure 11.5 may be the simplest Feynman diagram involved, but it is not the only one. To obtain a correct representation of the process *all* the relevant Feynman diagrams must be included. If we go to more complicated diagrams then we must include those that look like figure 11.4 as well (but with one of the electrons being an e^+). All the diagrams that we have drawn for e^-e^- reactions we can also draw for e^+e^-. However, there are some diagrams that can only be drawn for e^+e^-.

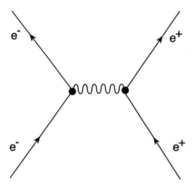

Figure 11.5. e^+e^- interaction.

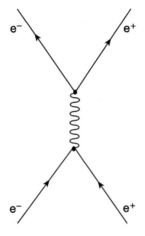

Figure 11.6. A distinct diagram for e^+e^- reactions.

Only when all the diagrams like figure 11.6 are included as well will the correct answer be obtained. The football exchange model does not work because it is wrong. At best it presents a glimmer of the true situation in one distinct case. It is better to surrender the need for a clear physical picture than to cause severe mis-representations.

Remember that no one Feynman diagram can ever tell the whole story.

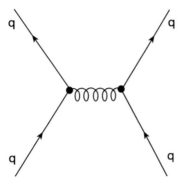

Figure 11.7. One gluon exchange.

11.2 Extending the idea

Such is the success of QED that theorists have copied the technique to try and understand the other forces. This programme has met with considerable success over the last 20 years. We now have fully functioning theories of the strong and weak forces along the same lines as QED.

11.2.1 QCD

Quantum chromodynamics (QCD) is the theory of the strong force. In outline it is the same as QED. If we wanted to calculate the probability of two quarks interacting we would start by drawing a Feynman diagram such as figure 11.7.

The differences in the theory emerge when we have to use more complicated diagrams. We discovered in chapter 8 that quarks come in three different 'charges' or colors. The whole theory of QCD is built around this idea (hence the name of the theory). One of the implications is that the gluons themselves must be colored (but note that in QED photons are *not* electrically charged). This means that gluons also feel the strong force. Hence we have to consider some strange looking diagrams.

Including such diagrams has a very interesting effect. The mathematical consequence is that the more complicated diagrams are *more* important

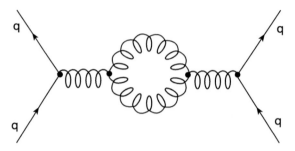

Figure 11.8. Gluon–gluon loops.

than the simple ones. Contrast this with QED in which the diagrams became less important as they increased in complexity. QCD is not a 'well behaved' theory and is very difficult to use in practice.

Including diagrams such as figure 11.8 also implies that the force between quarks *increases* with distance. This odd property of the strong force prevents us from observing isolated quarks in nature today[3].

As the quarks get closer together the force between them gets smaller. Hence as our experiments increase in energy the quarks inside hadrons will get closer to each other when they interact and so we will be able to get away with including fewer diagrams when we calculate to check the experimental results. Our approximations get better as our experiments increase with energy. However, some of the most important questions— such as the masses of the hadrons (see chapter 5)—cannot be answered at present. We await new mathematical or calculational techniques[4] in order to make progress.

11.2.2 Electroweak unification

One of the greatest successes of particle physics has been the development of a theory of the weak force. As an unexpected benefit the weak theory was found to be so similar to QED it became possible to develop a single mathematical theory that covered both forces. The complete theory is known as the *electroweak* theory. For this work Glashow, Salam and Weinberg received the Nobel prize in 1979.

Merging the electromagnetic and weak forces is an ambitious project. The two forces seem quite distinct from each other. The electromagnetic force is effectively infinite in range, whereas the weak force is confined to distances less than 10^{-18} m. The electromagnetic force only acts between charged objects, but the weak force can act between neutral objects (neutrinos react with quarks). Furthermore the weak force changes the nature of the particles!

If one tries to write down a mathematical theory of the weak force along similar lines to QED it turns out to be impossible unless you allow there to be *three* types of disturbance of the weak field. In chapter 7 we saw that the weak field has two charged disturbances called the W^+ and the W^-; now this theory is telling us that in addition there is a neutral Z^0 disturbance.

In the first versions of this theory the Ws and the Z were considered to be massless (just like the photon and the gluon) and so the Z seemed very similar to the photon. However, the simple theory failed as it could not explain why the weak force has such a short range.

The strong force is short range because the field energy increases as you separate the quarks (a consequence of the gluons carrying color) and the field decomposes into quarks when it gets too long (see chapter 5). This is not the case for the weak field. The theory suggests that the weak field behaves in the same way as the electromagnetic field—which has no limit on its range. The short range of the weak force must have a different explanation.

This problem held up the development of the weak theory until it was realized that an idea first developed by Peter Higgs in another context could be applied to the weak force. The price of the solution is to introduce another field into physics—the Higgs field. The Higgs field does not cause a force in the way that the other four fields do—it interacts with the weak field.

As W and Z particles pass through space they interact with the Higgs field and an exchange of energy takes place—the result of this being that the weak disturbances take on mass!

The whole subject of the mass of particles has been one of the outstanding problems of particle physics for 50 years. We discussed the problem of the mass of the proton in chapter 4. The Higgs field provides a way of explaining the masses of certain objects (Ws and Zs for example), but at best it is only a partial solution, as the values of the masses can still not be calculated from first principles.

As a way of imagining how the Higgs field provides mass consider an experiment in which a marble is fired from a gun inside a bath of treacle (the great thing about thought experiments is that you do not actually have to try to do them!).

Given a certain force on the marble we should achieve an acceleration in air. In the treacle the acceleration will seem to be less due to the drag of the treacle. If we were ignorant of the physics of friction in liquids we might explain this result by saying that the mass of the marble was greater in the treacle than outside it.

The Higgs field acts like the treacle on the weak force disturbances. Unfortunately the Higgs field is always present but cannot be seen, so we can never do an experiment without it to see the 'actual' massless nature of the weak disturbances.

When Higgs' idea was applied to the weak force (by Salam and Weinberg) an unexpected bonus was produced. The mathematical consequence of adding the Higgs field to the weak field was to 'split' the third disturbance (the Z^0) into two types—one the massive Z^0 and the other the photon. Electromagnetism was fully incorporated for free! The theory is also very specific about the way that the Z^0 interacts with other particles—it does not change the type of particle, but it can couple to objects with no charge. This produces an effect known as the 'neutral current' in which neutrinos can be seen to interact with other particles and stay as neutrinos. This type of reaction was discovered in 1973 and provided an early pointer to the theorists that they were heading in the right direction.

11.2.3 Experimental confirmation

In 1984 an experimental team at CERN lead by Carlo Rubbia announced the discovery of the W and Z particles. By colliding protons with antiprotons they had managed to muster enough energy to create the W and Z particles close enough to their correct mass to be produced in sufficient numbers to be measured in an experiment. The masses were measured and found to agree with the theoretical predictions. Rubbia and Van Der Meer were given the Nobel prize in 1984.

In 1990 the giant LEP accelerator started up at CERN. LEP collided e^+ and e^- at exactly the right energy to produce Z particles with their full mass. These then decayed in various distinct ways that could be seen in the experimental detectors, e.g.

$$e^+ + e^- \rightarrow Z^0 \rightarrow \mu^+ + \mu^-. \qquad (11.2)$$

LEP was a Z factory. Millions of Z particles have been created and measured. The agreement with theory is very impressive.

One of LEP's main aims was to search for Higgs particles—disturbances set up in the underlying Higgs field. Progressive improvements in the accelerator were made between 1989 (when it was first commissioned) and 2000. The original energy of 91.2 GeV would have allowed any Higgs particles with masses up to 65 GeV/c^2 to be seen. The subsequent upgrade to LEP2 peaked at 202 GeV in 1999. Still no Higgs particles were seen, pushing the lower limit on their mass up to 108 GeV/c^2.

In April 2000 a final push boosted LEPs energy as high as possible and 209 GeV was achieved—well beyond the original design limits. At last some indications of a Higgs particle with mass around 114–115 GeV/c^2 started to be seen by all four detectors linked to LEP. However, with so little data it is impossible to eliminate other reactions that may be taking place and producing a similar pattern in the detectors. The new information served to prolong LEP's life for a few weeks into October 2000, but then the decision to shut the accelerator down was taken so that work could start on the next design—the LHC. Here the situation remains. Analysis of the data that we have so far is still taking place, but it is likely that a formal announcement of the Higgs discovery will have to await the completion of the LHC in 2005. Fortunately, if the current mass estimates are correct, the Higgs particle should come flooding out of the LHC[5].

11.2.4   Counting generations

One of the most important definite results to have emerged from the LEP accelerator was the experimental confirmation that there are only three generations of leptons and quarks. This generation counting has been done by measuring the lifetime of the Z^0. The Z^0 is capable of decaying into any pair of objects that have less mass than it does. This obviously includes all the neutrinos. In broad terms, the greater the number of different decays that a particle has available to it, the shorter its lifetime is. The Z^0 lifetime has been measured and compared with calculations for two, three and four types of neutrino. The answer is clearly only consistent with three types of neutrino. From this we deduce that there are only three generations of leptons and, by symmetry, three quark generations as well. For the first time we can be sure that we have the complete list of elementary particles[6].

11.3   Quantum field theories

11.3.1   Exchange particles

We have now built up an impressively consistent view of how forces operate between particles. In each case there is a force field and the interaction between the particles gives rise to the disturbance in the field. This disturbance carries energy and momentum and in a Feynman diagram is represented as a particle moving between the vertices. The electromagnetic disturbance is the photon, the strong force has the gluon and the weak force the W and Z particles. The collective name for these disturbances is *exchange particles*.

All the exchange particles are in principle massless, but the Higgs field interacting with the Ws and Zs gives them very large masses.

The full masses of the Ws and Zs are, respectively, 86 GeV/c^2 and 93 GeV/c^2, which represents an enormous amount of energy to have to muster in the weak field.

Fortunately, we do not need all this energy to create the W and Z exchange particles. They can be emitted with very much less energy than this. When this happens they are said to be 'off mass shell', which is a very complicated sounding term but it just means that they have been emitted with less energy than they should have.

Now, an off mass shell particle is quite an odd thing. It certainly does not have the mass that we normally associate with the ordinary particle. It is a disturbance in the field that has been forced into existence rather than appearing naturally.

This is really an effect that can only be explained with quantum mechanics. The best analogy that I can come up with is to compare it to the sound made when you lightly tap a wine glass.

Every wine glass has its own natural tone, which it will produce when tapped. However, if we place a wine glass next to a loudspeaker and play a note very loudly, then the glass will try to vibrate with the same frequency as the tone being played. It will not be very successful and the vibration will die out rapidly after the tone stops (but if you tap a glass then the natural tone tends to last longer). We have forced the glass to vibrate at a frequency that is not natural. The price we pay is that the vibration does not last very long.

When two particles interact, the field between them is set into *forced vibration*. The result is the exchange particle that travels between them— but it is off mass shell. The closer the particles get to each other, the greater the energy in the field and the nearer to being on mass shell the exchange particle will be (the equivalent to the natural tone of the glass). This makes the production of the exchange particle much more likely.

This is why the weak force is short range. Trying to create Ws and Zs without enough energy available for their full mass makes them difficult to create unless the interacting particles are very close.

The photon and gluon do not suffer in this manner as they have no mass and so can be created relatively easily at any distance[7]. However, they can be off mass shell as well!

Consider colliding an electron and a positron. If they are moving with equal speed, then the their combined momentum will be zero. On colliding they will annihilate as in figure 11.6. The photon in figure 11.6 must have zero momentum, or momentum would not be conserved in the reaction. This would mean that it was stationary. However, photons normally move at the speed of light. This is no ordinary photon! This one is off mass shell—this one has *got* mass. Such a photon is almost literally pregnant with energy and must convert back into a particle–antiparticle pair rapidly (as in figure 11.6).

Exchange particles are always off mass shell to some degree. This is why they can only be found between the vertices of a Feynman diagram. Any particles entering or leaving the bubble of ignorance must be on mass shell.

If you like, you can imagine that off mass shell particles do not exist at all. After all, we can never observe them. No one Feynman diagram represents what is going on inside the bubble of ignorance. They each provide an approximate way of looking at the complicated disturbance of the field. Only in total do they give an exact description of the interaction. There is no single photon carrying all the energy in an annihilation such as figure 11.6. We have found that the only way to make Feynman diagrams work as descriptions of interactions is to use off mass shell particles in the diagrams. Particles entering and leaving the bubble of ignorance are ordinary, on mass shell, objects. Some people refer to off mass shell particles as *virtual particles*, a name that emphasizes their nature—virtual reality is a simulation of reality, not the real thing.

In chapter 3 I explained that an uncertainty principle exists between energy and time:

$$\Delta E \times \Delta t \geqslant h/2\pi$$

which is sometimes used to explain the existence of virtual particles. According to this view, a quark can 'borrow' the energy required to emit a W (for example) provided the energy is 'paid back' in time Δt. This means that the W has time Δt to get from one quark to another before it has to disappear again. Consequently as the W is a heavy particle ΔE must be large making Δt small and the force is therefore short range.

I do not find this a very helpful way of thinking about virtual particles as it can lead to questions such as:

• How does the quark 'know' how much energy to 'borrow' before it tries?
• What would happen if the time ran out before the W got there?

Of course the uncertainty principle does apply to particle interactions, but it is related to the bubble of ignorance in figure 11.1. The interaction taking place inside this bubble will last for a certain duration Δt. This implies that the amplitude for the total energy involved in the interaction is spread over a range of energies $\Delta E = h/2\pi \Delta t$. The shorter the duration of the interaction is, the wider the range of energies that could be involved (and vice versa). The uncertainly principle is expressed in the range of Feynman diagrams that are being considered, rather than within a given diagram.

11.3.2 Quantum fields

The world contains various force fields with which particles interact. Interactions cause energy to be exchanged between the particles and the fields and the result is an exchange particle, which is a moving disturbance of the field.

It is an interesting exercise to think about the difference between an ordinary particle, say an electron, and one of these exchange particles. The full mathematical theory of quantum fields suggests that there is not much difference! Just as there is an electromagnetic field with its disturbance, the photon, it is perfectly reasonable to think of an electron field containing a disturbance that we regard as a moving electron. At first this sounds rather odd. After all, we are used to thinking about electrons as little lumps of matter. However, if you think back to quantum theory in chapter 3 then you should be prepared for anything!

This is actually a beautiful generalization of our thinking. The electron field is an object that is bound by the requirements of relativity (hence it is called a *relativistic quantum field*), so it actually describes both electrons and positrons. Consider again figure 11.6. There is another way of thinking about the process represented by this diagram. The electron and positron are both excitations of an underlying quantum field. When they annihilate the energy is dumped into the electromagnetic field forming a photon. This energy does not remain in the electromagnetic field but is dumped back into the electron field (to which it is coupled) producing a new electron–positron pair.

The quantum mechanical description of a quantum field is in terms of amplitudes that represent the probability of a field containing various excitations. For example the amplitude $\phi(n_1, n_2, n_3, n_4, \ldots, n_i)$ governs the probability that the field contains n_1 particles of momentum

p_1, n_2 particles of momentum p_2 etc where n is a whole number
(including zero)[8]. A quantum field containing just the one excitation
would reduce to the sort of amplitude that we became used to in
chapter 3. A field that contained two particles interacting with
each other could be $\phi(1, 0, 0, 0, 1, 0, \ldots, 0)$ before the interaction
and $\phi(0, 1, 0, 1, 0, 0, \ldots, 0)$ after it (provided momentum has been
conserved—in other words I have got the 1s in the right places!).

These quantum field amplitudes are generalizations of the amplitudes
that we called wave functions in chapter 3. You will recall that
Feynman's sum over paths allows physicists to calculate one wave
function given another. The expression I quoted in chapter 3 was

$$\varphi(x_b, t_b) = \int K[(x_b, t_b), (x_a, t_a)]\varphi(x_a, t_a)\, dx_a$$

where the funny looking K thing was the amplitude for a particle to get
to (x_b, t_b) from a given starting point by covering every possible path.
The calculation involves taking this amplitude and using it to get the
wave function at (x_b, t_b) by adding up all the amplitudes to get to that
point from any starting point (that is what the integral over x_a is doing).

In the quantum field picture we want to calculate the amplitude for the
field to get from one state to another:

$$\phi(n_1, n_2, n_3, n_4, \ldots, n_i) \rightarrow \phi(m_1, m_2, m_3, m_4, \ldots, m_i)$$

where n and m do not have to be the same numbers, which is just like
getting from one wave function to another. The basic principle is the
same, but now there are multiple fields involved with various interactions
between them so the method becomes one of summing over all the
intermediate states and interactions—this is the 'path' between the two
field configurations. It is from this method that the Feynman diagrams
emerge. They are the different 'paths' as represented by K.

Our picture of reality now looks something like this. The universe is
full of various quantum fields that describe the elementary particles and
the fields of force[9]. These fields can couple with one another causing
disturbances to emerge and disappear back into the underlying quantum
field. These are the various interactions that we see happening between
'particles'. Each represents a possible path between different states and,

according to Feynman, can be associated with an amplitude. To get the correct final amplitude for a change of field configuration each of these amplitudes must be summed.

11.3.3 Spin

Spin is a property shared by all particles. In essence spin is a quantum mechanical property and there is nothing like it in classical physics (where have we heard that before...).

To a limited extent one can imagine that every type of particle spins on its axis, like a top, as it moves through space. The speed at which it spins is fixed for each particle. The units that we use to measure the rate of spin are in multiples of Planck's constant $h/2\pi$ (often abbreviated to $\hbar$). Quarks and leptons all spin at the same rate of $(1/2)\hbar$, photons, gluons and the other field disturbances have a spin of exactly $\hbar$. Even neutrinos spin, although the antineutrino spins in the opposite direction to the neutrino.

If the quarks have spin, then the hadrons made from them also must have some spin. A complex set of rules governs how the spins of the individual quarks combine to produce the spin of the hadron.

When they are in atomic energy levels electrons are allowed to spin on their axes as well as orbiting the nucleus. The direction of the spin can either be in the same sense as the orbital rotation or opposite to it (i.e. both clockwise or one clockwise and one anticlockwise)—the electrons are not allowed to 'roll' round in their orbits. A similar situation applies to quarks inside hadrons. When there are three quarks in a baryon the rules dictate that at least two quarks must be spinning in the same state. Consequently the total spin can be either $(1/2)\hbar$ (two in one sense and one opposite) in the case of the baryon octet, or $(3/2)\hbar$ (all quarks spinning in the same sense) in the case of the baryon decuplet. Protons and neutrons have $(1/2)\hbar$ spin, the Δ baryons are spin $(3/2)\hbar$.

In a meson the quark and antiquark spin in the opposite sense so the overall spin of the composite particle is zero. Particles that spin with a 1/2 multiple of $\hbar$ are called *fermions* (i.e. leptons, quarks and baryons) those that spin with a whole number multiple of $\hbar$ are called *bosons* (i.e. photons, gluons etc).

In section 6.4 I pointed out that the quarks occupying energy levels inside hadrons are subject to a rule similar to that which governs the arrangements of electrons in atomic energy levels. This is known as the *Pauli exclusion principle* (after its discoverer—the same Pauli who suggested that neutrinos existed) and it applies only to fermions. Basically the rule states that no two fermions of the same type (two electrons or two quarks) can be in the same quantum state. Consequently in a quantum field that describes fermions —$\phi_f(n_1, n_2, n_3, n_4, \ldots, n_i)$ the ns can only be 1 or 0. The same restriction does not apply to boson quantum fields. In that case there can be any number of excitations of a given momentum. This has physical consequences as well. A large number of bosons in a similar state leads directly to the sort of electromagnetic and gravitational fields that we experience in the macroscopic world. Excitations in a fermion field cannot act in concert in the same manner, which is why we do not see their field nature macroscopically—they look like individual particles.

This means that there is some vestige of our old fashioned distinction between force and particle remaining—the quantum fields of forces are bosonic and the quantum fields of particles are fermion fields.

11.3.4   Renormalization

QED is a highly successful theory, but it had some technical difficulties in the early days of its development. One of these was the unfortunate tendency to come up with answers that were mathematically nonsensical.

This issue centres round so called 'loop diagrams' like the one in figure 11.9. Feynman's prescription for doing quantum mechanics demands that all possible diagrams must be taken into account, so odd things like this have to be included as well.

The problem lies with the electron–positron loop at the centre of figure 11.9. The photon propagating across the diagram is moving through a background containing various quantum fields (which are not shown on the diagram just as the background electromagnetic field is missing). In this case the photon's coupling to the background electron field has caused an electron–positron pair to materialize out of the background. Such a process is known as a *vacuum fluctuation*.

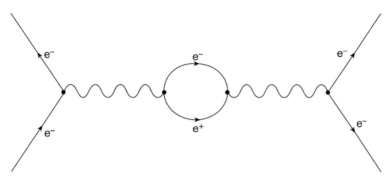

Figure 11.9. A loop diagram that contributes to electron–electron repulsion. In this case an electron and a positron have been created in the loop—but it could be a pair of muons etc.

The electron and positron that materialize, and then de-materialize again, must be off mass shell. Unfortunately as they form a closed loop there is no way of telling what their energy and momentum must be—it is not determined by any of the other features in the diagram. Consequently, when the calculation is done, this diagram has to be included an infinite number of times—one for every possible energy–momentum combination in the electron–positron loop. This leads to the whole calculation 'blowing up' and producing an infinite answer.

There are two steps to solving this problem. The first is to carry out the calculation corresponding to this diagram, but to stop at a 'cut-off' energy. This is justified by assuming that at high energy QED ought to be replaced by a better (unknown) theory anyway! Doing this produces an answer that depends on the electron's charge (this comes in at each vertex—the black blobs) and on the value of the cut-off energy. The arrangement of terms in this answer makes it possible to combine the charge and the factor due to the cut-off energy into one part:

$$e_R = e + \delta e$$

where e is the electron's charge and δe is the part of the equation that depends on the cut-off energy—this piece is being interpreted as a modification to the charge on the electron! However, all is not totally well, as the second piece still depends on the cut-off energy, and this term will still blow up if we allow this cut-off to go to infinity, as it should.

The next part of the trick is to think carefully about what really needs to go into the calculation. The electron charge, symbolized by e in the equation above, is never actually measured. More to the point it never can be. It is know as the *bare charge*. Any process that measures the charge on an electron *must* involve interacting with the electron and so must result in an exchange of off mass shell photons—just as in the Feynman diagrams being considered. Consequently, in any experiment that sets out to measure the charge on the electron, loop diagrams are going to have to take part. The upshot of this is that you can only ever measure the modified charge on the electron—e_R not e.

By now the reader will probably be protesting that this is absurd, as we have not measured the electron's charge as being infinitely large. Quite true, e_R is a finite value $(1.6 \times 10^{-19}\,\text{C})$ *so this is what we should put into our calculations that use e_R.* In other words we know that e_R is a finite value so we put the *measured value* into the problem and stop trying to calculate δe. The trick is called *renormalization*.

If you are still not happy, then try thinking about it like this—say that the bare charge is split into two pieces e_1 and e_2 so that $e = e_1 + e_2$, then e_R becomes:

$$e_R = (e_1 + e_2) + \delta e.$$

Now, if e_2 was infinite but of the opposite sign to δe, then those two would cancel out and we would be left with the finite part e_1 which is obviously the same as e_R. Renormalization is effectively assuming that something like this is going to happen.

If this sounds like various ways of trying to justify sweeping a problem under the carpet, the reason is that is exactly what is happening. The miraculous thing is that it works! The electron has the same charge as the other leptons and the diagrams that they are involved with can be 'corrected' in just the same manner. The problems can all be swept under the same carpet using the same brush.

The process is not completely over. The final step is to replace the expression for e_R in the calculation with just the numerical value and to see what happens (we no longer have to worry about the cut-off and taking it to infinity as that is presumed to be done resulting, somehow, in the finite value e_R that we measure). When this is done, an interesting thing emerges. There is still another factor to do with the loop diagram remaining—a factor that depends on the energy of the photon, q. This factor can now be combined with e_R in a similar manner to the way in which the cut-off term was, producing another modification this time to e_R:

$$e_R(q^2) = e_R + \Delta(q^2)$$

where q is the energy of the photon crossing the diagram[10]. The modification term $\Delta(q^2)$ is finite and perfectly well-behaved mathematically, but it does say something rather remarkable—that the charge on an electron is a function of energy.

One way of thinking about this is to remember that there is no such thing as an electron that is genuinely on its own. Any electron is part of the underlying electron field and, as such, when a photon approaches it there is always the possibility of loop interactions forming. We interpret this by saying that the electron's charge changes with the interaction energy, but really we are saying that the higher the energy the more complicated looking the electron is. This has some directly observable effects, the most celebrated of which is the 'Lamb shift', a small variation in the frequency of light emitted by electrons in hydrogen atoms. This deviation from the expected values can be explained by including the vacuum fluctuation diagrams similar to those in figure 11.9.

Loop diagrams also exist in QCD such as that in figure 11.8. These would also blow up if the effects were not incorporated into the renormalized color charge of the quarks. This also produces a color charge that depends on energy, but in the opposite sense: the higher the

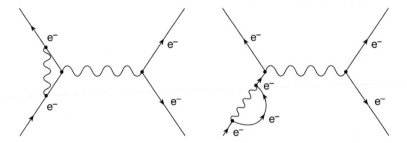

Figure 11.10. Some other loop diagrams. The diagram on the left will contribute to the renormalization of the electron's magnetic properties. The one on the right will be swept up in the renormalization of the electron's mass.

energy the smaller the apparent color charge. This is another reason why it is much easier to deal with QCD at high energies.

QED also allows other loop diagrams that must be dealt with. Fortunately a powerful mathematical theorem shows that all the loop diagrams that exist can be characterized into types and swept up into the renormalization of other physical quantities. Examples of this are given in figure 11.10.

11.3.5   The vacuum

It must by now have become clear that the quantum relativistic description of the world is a very strange one indeed. Unfortunately we must now add a further aspect to it that will make it seem utterly bizarre, but which may, nevertheless, have extreme significance for the whole universe.

The sort of loop diagrams considered in the previous section were referred to as vacuum fluctuations. The name comes from the apparent ability of the photon in figure 11.9 to make an electron–positron pair out of nothing—the vacuum. Vacuum physics is much more complicated than one might have thought. Even if there is a volume of space totally empty of particles that does not mean that the underlying fields are absent. If there are fields present, then perhaps there is energy present as well. That being the case, we must allow for more vacuum fluctuations to take place—even if there are no other interacting particles present.

In figure 11.11 an electron–positron pair has materialized out of the vacuum. They are both highly off mass shell so cannot be observed directly, unless they happened to interact with some passing particle, in which case they would become part of a loop diagram. The energy required has been donated from the store present in the vacuum and returned again when the two particles disappear. The vacuum is a seething mass of such materializations driven by the store of energy present. Unfortunately this vacuum energy (or zero point energy as it is sometimes known) cannot be calculated. It is a constant background factor in all our calculations, but much like the bare charge it can never be directly observed[11].

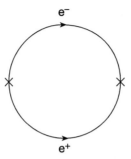

Figure 11.11. A vacuum fluctuation. In this diagram I have used crosses instead of blobs at the vertex to symbolize that the coupling is to the vacuum rather than to any other particles.

It is a little bit like the depth of a lake. Once the water is more than a certain depth it has no effect on the way in which ripples move over the surface. Any amount of study and calculation about the way in which ripples cross the lake and interact with each other will not reveal anything about the depth of the lake.

This is the situation we face with relativistic quantum fields. The interactions that we study are 'ripples' on the surface of the 'lake'. When there are no ripples present, we used to think that there was no lake either—a 'true vacuum'. Now we realize that the particles we see are ripples on the surface of a vast 'lake'—the quantum field—and our equations do not really give us a handle on the way to calculate how much energy there is in this field[12].

Some estimations of the vacuum energy have been made producing extraordinary results like 10^{85} J in every cm^3!

As we will see in chapter 15 the very latest results from observations of distant objects suggest that the expansion of our universe is accelerating, not slowing down as one might expect due to the action of gravity. Einstein noted the possibility of this in his theory of general relativity, which can produce a form of repulsive gravity given a particular form of background energy in the universe. The vacuum energy of quantum fields is just the right form of energy to do this. Unfortunately, the estimations of how big this background energy can be (from astronomical data) have come out as 10^{120} times *less* than that estimated from quantum field theory. This has been called the worse back of an envelope estimation ever made.

If the vacuum energy is producing an accelerating expansion, this would be another extraordinary link between the microworld and the physics of the whole universe. The story will be taken up further in chapter 15.

11.4   Grand unification

Progress in science is not simply driven by experimental success. The feelings and intuitions of the people involved also have an active part to play. In physics there has been a long held aesthetic desire to see simplicity. In the history of the progress of the subject great advances have been made on the basis of choosing the simplest and most mathematically elegant option. The success of electroweak unification has fuelled the aesthetic need for a single theory that unifies all the fundamental forces. As a result this has been an active area of research for a number of years.

Vacuum fluctuations get in the way of direct interactions between bare particles and as a result the 'charges' dictating the force's strength change with energy. In a remarkable series of calculations a group of physicists[13] showed that the effects of these vacuum fluctuations converge at high energy. In other words as the energy of the interaction gets strong enough, so the strength of the three non-gravitational forces become the same! Perhaps there is only one charge governing the strong, weak and electromagnetic forces and what we see as three different effects is due to the masking of this charge by different vacuum fluctuations.

Figure 11.12 shows qualitatively the results of these calculations.

Note that the weak force appears to be *stronger* than the electromagnetic force on this diagram. This is because the graph is showing the 'inherent' strength of the force, which takes the vacuum fluctuations into account, but not the mass of the W and Z particles. As they are so massive, they are difficult to emit and this reduces the 'real' strength of the force to less than that of the electromagnetic force.

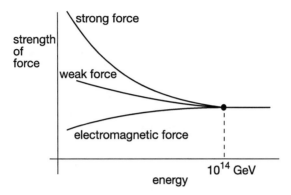

Figure 11.12. The variation of force with energy.

All three forces become comparable in strength at the stunning energy of 10^{14} GeV. This is well beyond the range of any conceivable particle accelerator. However, this does not stop theorists from speculating about the nature of a 'grand unified theory' (GUT) in which the strong, weak and electromagnetic forces are seen as the disturbances of a set of basic fields linked via the Higgs field. As a consequence there should be new fields that have not yet been seen with their own disturbances, the X and Y particles. The detailed properties of these particles depend on the exact version of the theory used, but their masses must be about 10^{14} GeV/c^2.

Although we do not know which GUT is correct at the moment, they all predict that protons should not be stable particles. The quarks inside should be able to emit X and Y particles and turn into leptons (hence the baryon number is not truly conserved), but we are such a long way off the GUT energy that these processes must be very unlikely. As a consequence the lifetime of the proton is predicted to be of the order of 10^{30} years!

The greatest hope that the theorists have of testing their theories lies in speculating about the early universe. We believe that shortly after the big bang all the particles had energies above that required by the GUT theories—perhaps the nature of the interactions that went on then will provide some fingerprint on the universe that can still be seen today. We shall take up this idea in chapter 11.

11.5   Exotic theories

A promising modern theory attempts to link fermions and bosons and would allow a particle to convert from one type to another. Supersymmetry (SUSY) is a very elegant theory that is widely held to be true (at least in some respects) in particle physics circles. Unfortunately there is no experimental evidence for the particles that it predicts at present (although there have been some hints from the most modern results).

Supersymmetry was first discussed in the 1970s as an academic exercise. Theoreticians explored the mathematical structure of a theory that made no distinction between fermions and bosons, in the sense that the particles could be swapped round without altering the theory. They discovered that if the equations were to work they had to include a reference to a field that had not been originally specified in the theory. The properties of this field were strongly reminiscent of the gravitational field as described by Einstein's theory of general relativity. This exciting result held the first hint that it might be possible to draw gravity into the realm of particle physics.

However, there was a price to pay. The theory implied that there must be a fermion equivalent to every boson and vice versa. The spin-1 equivalent to a quark is called a squark and that of a lepton a slepton. Along with this are the photinos (spin-1/2 photons) and the winos (spin-1/2 Ws). Clearly this theory has allowed physicists' fluency with names to blossom. At a stroke the number of elementary particles has been doubled. Theorists counter this criticism by pointing out how neatly the theory manages to deal with the problems inherent in gravitational theories. At the moment, the idea is viewed very positively. However, the supersymmetric particles have never been observed, implying that they are more massive than can be produced in our current accelerators. The theory requires some doctoring for the supersymmetric particles to have different masses from their common counterparts. The next generation of accelerators should settle the issue.

There is, however, one interesting piece of evidence that is already to hand. If you recall figure 11.12 it shows the interaction strengths of the non-gravitational forces coming together at a high enough energy. This calculation was re-done[14] in 1991 using more modern measurements of the force strengths at more conventional energies. The result showed that the forces converged to being very close to each other—but missed being exactly the same. However, if the supersymmetric particles are allowed to be part of the calculation as well (their vacuum fluctuations are included), then the forces do converge to exactly the same strength!

11.6 Final thoughts

And so to even more speculative ideas.

If the principle of GUTs is correct, then we still have gravity to include in our scheme. At some time after the big bang the energies involved were such that gravity was comparable in size to the GUT forces. When this happens physics really does enter wonderland. The energy at which this is expected to happen is approximately 10^{40} GeV and all of our common-sense notions of space and time are rendered useless. This is an exciting area of physics. It is also the hardest to grasp mathematically and conceptually. The most promising line of attack seems to be superstring theory.

A superstring is a little closed loop of spacetime. When the ideas of quantum theory and supersymmetry are used to frame the structure of such an object, then a surprising feature emerges. The strings are able to 'vibrate' at various frequencies. Each mode of vibration gives the superstring the properties of a fundamental particle—in one case a quark, in another a lepton then an exchange particle etc. By allowing the strings to interact with each other all the fundamental forces, including gravity, can be incorporated into one system. Technically this is a stunning success as the finite size of the strings puts a natural limit on the energy of the exchange and disposes of the need to renormalize the forces. However, all this success has to be bought at a price. The mathematical machinery of the theory is extremely complicated and in some instances has yet to be fully worked out (there was a nice comment made a few years ago that superstring theory was a bit of 21st century physics that happened to drop into the 20th century by mistake...). Also the theory has to be couched in more than four dimensions (10, 13 or some other higher number) in order to give the strings the necessary freedom. Naturally the other dimensions have to be suppressed in some manner so that we live in a four-dimensional (three space and one of time) universe. The model employed for this is that the other dimensions are closed (like the whole universe is closed as described in chapter 13) on themselves in a tiny region about 10^{-33} m across—consequently if you set off along this direction then you would be back to your starting point within 10^{-33} m.

Superstring theory is still at a comparatively early stage of development, but progress has been encouraging. The real problem is trying to get some experimentally testable predictions from the theory, otherwise it is a purely mathematical exercise. The best hope for this lies in the application of string theory to the early universe when conditions were such that the energy was high enough to make the string nature of reality clear.

11.7   Summary of chapter 11

- Our understanding of forces comes from calculating the probabilities of various particle reactions;
- we accept that we can never know exactly what happens in a given situation, but instead we have to consider all the possible interactions that give rise to the same result and add up their probabilities;
- QED was the first theory to carry this programme out completely;
- Feynman's technique was to calculate the contribution of each interaction by listing all the possibilities in the form of diagrams;
- each diagram represents in a pictorial way a term in a mathematical approximation—it does not represent 'what happens';
- in QED the diagrams become less important as they become more complicated;
- QCD is the equivalent theory for the strong force;
- QCD is more difficult than QED as the diagrams become more important as they increase in complexity;
- the strong force gets weaker as energy increases, so it becomes possible to use approximation techniques at high energy;
- the weak force can only be dealt with by introducing a new field—the Higgs field;
- the Higgs field gives the W particles mass and 'splits' the third weak particle into the Z^0 and the photon—hence QED is incorporated into the weak theory;
- the W and Z particles have large masses, which makes the force weak and short range;

- GUT theories predict that the proton will decay and that the baryon number is not conserved;
- the only way of testing GUT and more exotic theories is to see what they say about the early universe.

Notes

[1] What makes the difference here is that it is impossible to interact with a subatomic particle in a *small way*. We can only interact with them via other subatomic particles, and that takes place via the sort of reaction that we are trying to study!

[2] Compare this to the following situation. It is well known that π is approximately 22/7; it is actually the sum of the following sequence: $\pi = 4(1 - 1/3 + 1/5 - 1/7 + 1/9 - \cdots)$. If you only use a few of these terms, then you will get an aproximate value for π. In the same way, if you only use a few Feynman diagrams you get an approximate view of the interaction.

[3] However, as we shall discuss in chapter 11 in the early phases of the universe all the quarks were so close together that the forces between them were zero and they acted like separate objects.

[4] Parallel processing computers have been developed partly under pressure from particle physicists to help with QCD calculations.

[5] The situation of August 2001 is that the suggested Higgs mass is 115.6 GeV/c^2, but there is a 3% chance that the data have been produced by other physics.

[6] Perhaps I ought to stand back from that statement a little. If there are any neutrinos with masses greater than the Z^0 then this technique would not count them. This, however, seems very unlikely. It is possible that there might be smaller objects inside both quarks and leptons, but we have managed to count the constituents of matter at this level.

[7] The reason why the strong force is short range is that when enough energy is built up in the strong field a quark–antiquark pair tends to pop up rather than a gluon.

[8] As field theory is all about interactions, it is most convenient to deal with the field by describing it in terms of states of particles with reasonably definite momentum—as we saw in chapter 3 this means that they have no well defined positions.

[9] The quarks and leptons have quantum fields—similar fields can describe the hadrons, but they should reduce to combinations of quark and strong fields.

[10] Strictly q is the 4-momentum of the photon, so q^2 is related to the mass of the virtual photon.

[11] Another way of thinking about this is to realize that the vacuum is another quantum field state. As such it is subject to energy/time uncertainty. If the vacuum is probed for a short duration, then it is in a mized state which includes a wide range of energy levels.

[12] Another analogy would be a bank balance. The monthly ingoings and ougoings of the bank balance are not influenced by how much is present in the balance. Interest payments sometimes go in and out—but this is a bit like interacting with the 'vacuum', the constant background state of the balance.

[13] David Gross, Frank Wilczek and (independently) David Politzer first showed that the vacuum fluctuations made the strong and weak forces weaker as the energy increased. Taking this idea up Howard Georgi, Helen Quinn and Steven Weinberg showed that the force strengths converged to the same value.

[14] Ugo Amaldi, Wim de Boer, Herman Fürstenau.

Interlude 2

Antihydrogen

Physicists believe that matter and antimatter behave in exactly the same manner in most circumstances. Certainly antimatter particles do not have 'negative mass' or antigravity properties and so they are not the answer to many propulsion problems as is assumed in some of the worst science fiction.

However, it would be sensible science to check this and so there are many experiments going on at the moment to see if matter and antimatter particles have the same mass. Unfortunately, it is very difficult to experiment with antimatter. If an antimatter particle comes into contact with a matter particle of the same type, then they will annihilate each other into energy. The amount of energy released in a single reaction of this sort is tiny and poses no safety risk—the problem is keeping the antimatter particles away from the matter particles for long enough to experiment on them (remember any equipment used will be made of the matter particles!).

At CERN they have partially solved this problem by building the Low Energy Antiproton Ring (LEAR). LEAR can store samples of antiprotons by circulating them back and forth in a 20 m diameter ring for several days. The antiprotons circulate in a small vacuum pipe and are kept in the centre (well away from the walls) by bending magnets. At several points round the ring there are detectors that monitor the position of the antiprotons. If they start to move out of place a signal is sent across the diameter of the ring to focusing magnets on the other side. By the time the antiprotons arrive the magnets are ready to squeeze them back into place. The developer of this technique, Simon Van der Meer, was awarded a share of the Nobel Prize in 1985.

In September of 1995 Professor Walter Oelert and his team from several European universities announced that they had used antiprotons from LEAR to successfully manufacture atoms of antihydrogen (experiment PS210). This was the first time that a complete anti-atom had been made. Antihydrogen consists of an antiproton with a positron in orbit round it. Physicists were quite sure that such objects should exist—indeed it should be possible to make anti-versions of all the chemical elements—but the experiment was a welcome confirmation of their ideas.

The experiment worked by allowing antiprotons from LEAR to pass through a small jet of xenon gas (see figure I2.1). Occasionally an antiproton passing close to a xenon nucleus would interact with its electromagnetic field and emit a photon. This photon might then convert into an electron–positron pair.

Although it is possible for any photon to do this, the electron and positron are normally a long way off mass shell as the photon does not contain enough energy to genuinely materialize this pair of particles. This would mean that the e^+ and e^- re-annihilate very soon after they are created. However, it is possible that during the short time in which they exist one of them could absorb a photon emitted from the xenon nucleus. If one of them were already close to being on mass shell, and the other, which was a long way off, were the one to absorb the photon, then it could provide the extra energy needed to supply its full mass. This would then mean that both the e^+ and the e^- could exist indefinitely.

The chain of circumstances needs to go one link further to create antihydrogen. The positron has to be moving slowly enough to be captured by the antiproton as it passes.

Clearly the whole sequence of events is rather unlikely (the probability has been estimated as 10^{-17}!), but it does happen. In the course of three weeks antiprotons circled the LEAR ring and passed through the xenon 3 million times a second. *Nine* antihydrogen atoms were produced.

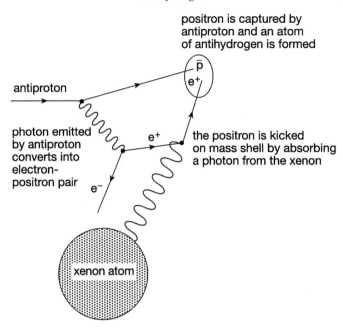

Figure I2.1. The chain of events leading to an atom of antihydrogen.

Once an atom of antihydrogen is created it is neutral and so it is not affected by the bending magnets in the LEAR ring. The anti-atom flies off from the ring and can be detected with suitable equipment.

This experiment represents a first step in the study of anti-atoms. The antihydrogens produced only lasted for about 40 billionths of a second. This is far too short a time to enable anything other than a confirmation of their existence to be done. The next step is to try to trap some of these anti-atoms so that their properties can be studied in detail.

Chapter 12

The big bang

In this chapter we shall explore the evidence for the big bang theory of the creation of the universe. We shall also examine the implications of particle physics for this theory and hence see why the physics of the microworld has important implications for the universe as a whole.

12.1 Evidence

'It is a capital mistake to theorise before you have all the evidence. It biases the judgement' — *Sherlock Holmes* From *A Study in Scarlet* by Sir Arthur Conan Doyle.

Today, most astronomers and astrophysicists believe that the universe was created in an extraordinary event, referred to as the big bang. Even though this took place some 15 billion years ago, we can be confident that our belief is sound due to three pieces of observational evidence that fit neatly into the scenario of a big bang creation. Other explanations for these observations have been offered, but none of them has the simplicity or inter-relatedness of the big bang idea. We shall look at each piece of evidence in turn and then examine how the big bang model accounts for them.

12.1.1 Redshift

In 1929 Edwin P Hubble published the results of a series of observations made with the 100 inch diameter reflecting telescope at Mount Wilson, near Los Angeles. This work extended the earlier contribution of Vesto Melvin Slipher who in 1914 had shown that the light from several galaxies was shifted towards the red end of the spectrum. Hubble examined the light from more galaxies and also managed to estimate the distances to both his and Slipher's galaxies. As a result he was able to show that the redshift increased with distance—'a roughly linear relationship'. This has become known as *Hubble's law* and its discovery marks the start of modern observational cosmology.

Stars and galaxies emit a wide range of wavelengths in the electromagnetic spectrum. In the central cores of stars, the atoms are interacting with one another to such an extent that the normal line spectra of isolated atoms are merged into a continuous band[1]. This light has to pass through the outer layers of the star to reach us. The outer layers are less dense (hence the atoms act individually as in a gas) and much colder than the core so the atoms will be tending to absorb light rather than emitting it. Hence, the continuous spectrum of light produced in the core of the star is modified by having dark lines cut into it where atoms in the outer layers have absorbed some wavelengths.

By looking at the patterns of wavelengths that are absorbed, astronomers can tell which atoms are present in the outer layers of a star.

These dark lines, called Fraunhofer lines, after the German optician who discovered them in the sun's spectrum[2], allow us to measure the extent to which a spectrum has been shifted. Ordinarily, if a spectrum is shifted towards the red wavelengths, then the longest wavelengths become longer—i.e. they become infrared and hence invisible, but at the other end the ultraviolet wavelengths become violet and hence visible. The visible part of the spectrum is therefore unchanged (see figure 12.1).

The Fraunhofer lines are produced in unique wavelength patterns characteristic of each atom. If they are shifted, the astronomer can see by how much they have been moved along the spectrum (see figure 12.2). Without the Fraunhofer lines we would not be able to tell that a continuous spectrum had been shifted.

Given the wavelength, λ, of a line within the spectrum and the measured wavelength, λ', of that line in the light from a galaxy, then we define the

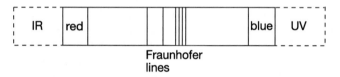

If the spectrum is shifted towards the red, then the UV light
becomes blue and the whole spectrum is unchanged

Figure 12.1. The shift of a continuous spectrum cannot be seen (the dotted
region marks the invisible wavelengths).

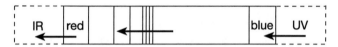

If the spectrum contains Fraunhofer lines, then
they will also move away from their correct wavelengths

Figure 12.2. The Fraunhofer lines provide a 'marker' by which shift can be
measured.

redshift of the galaxy as being:

$$z = \frac{\lambda' - \lambda}{\lambda}.$$

Hubble's observations suggest that the size of the redshift, z, is
proportional to the distance to the galaxy, d:

$$z = H'd.$$

In this equation, H' is related to a very important constant in modern
cosmology—the *Hubble constant H*:

$$H' = H/c$$

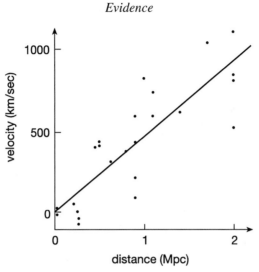

Figure 12.3. Hubble's original data. The horizontal scale is plotted in megaparsecs (millions of parsecs), the parsec being an astronomical unit of distance roughly equal to three light years. The vertical axis is plotted as speed rather than redshift. This can be done if the redshift is interpreted as being due to a Doppler shift in the light of a moving galaxy (see section 12.3). z is related to the speed by $z =$ velocity of galaxy$/c$.

or

➡ $$H = H' \times c.$$

An accurate measurement of the Hubble constant would help us to pin down the current age of the universe as well as its ultimate fate (as explained later). However, H is notoriously difficult to measure (basically the galaxies are a very long way away, so rather indirect means have to be employed to estimate the distances involved). However, techniques are steadily improving and the data are certainly far more convincing now than it was when Hubble first published his results.

Modern astronomers are deeply suspicious of any suggestion that our position in the universe is a special one. Why should the redshift of distant galaxies be proportional to their distance from our galaxy? The big bang theory provides a satisfying answer to this question as well as explaining the physical mechanism for the redshift.

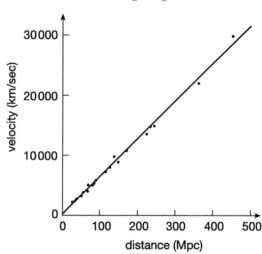

Figure 12.4. Modern evidence for Hubble's law. This much more modern plot (1996) shows how we have been able to push Hubble's measurements to far greater distances (compare the horizontal scales!).

12.1.2 Helium

The universe as we observe it today consists of about 70–80% hydrogen and 20–30% helium. All the elements heavier than helium, like the carbon from which we are made, are trace impurities added to this basic mix. Any theory of the creation of the universe has to explain the observed abundance of the elements. In 1957 a famous paper by Burbidge, Fowler and Hoyle showed that the atoms heavier than helium could be built up by nuclear fusion taking place inside stars and during the violent supernova explosions that can destroy certain types of older stars. Since then astrophysicists have become convinced that all the heavier elements have been made by the fusing of lighter elements inside stars[3]. However, the observed abundance of helium cannot be explained in this way.

There are two basic reasons for this.

Firstly, the amount of helium that we can see inside stars (by looking at the Fraunhofer lines) seems to be independent of the age of the star. If the helium content of the universe has been built up over long periods, then one would expect the older stars to contain less helium than the younger ones, which have been made out of more recent material. No such variation is observed.

Secondly, the process of forming helium out of hydrogen releases a great deal of energy (this, after all, is how a hydrogen bomb works). The release of energy required to build up 25% helium in the universe would make stars shine far more brightly than they do.

Hence we require some other mechanism to explain why 25% of the mass of the universe is helium. Furthermore this mechanism must have produced all this helium *before* the oldest stars were formed.

12.1.3   The cosmic background radiation

In 1965 Robert Wilson and Arno Penzias published an accidental discovery that revolutionized our thinking about cosmology. For a couple of years they had been using a radio antenna (built by Bell Laboratories in America for satellite communication) to study radio emissions from space and had found a low intensity signal in the microwave region of the electromagnetic spectrum. The signal appeared to be coming equally from all directions. Normally a radio source in space will be localized to a particular part of the sky. For example, the planet Jupiter in our own solar system is a very strong source of radio waves due to the constant thunderstorms that are taking place in its atmosphere. Certain types of galaxy are very strong emitters of radio waves and their noise can be identified as coming from a particular part of the sky. The hiss that Penzias and Wilson had picked up was the same intensity in every direction and at any time of the day or night. For this reason they suspected that it was interference produced within the antenna itself. However, after a careful study and inspection of the device (including the eviction of some pigeons that had taken to nesting within the antenna) they could find no explanation for the noise. Eventually they were forced to conclude that it was coming from outer space.

Since the initial discovery in the centimetre wavelength range, measurements of this background radio noise have been made at a variety of other wavelengths. The results of these measurements have shown that the intensity of the noise varies with wavelength in a characteristic way well known to physicists—a black body spectrum.

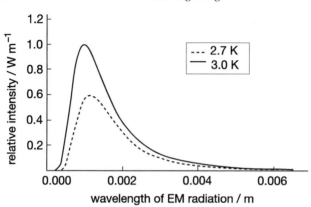

Figure 12.5. A black body spectrum.

All objects that are in equilibrium with the radiation they produce emit a characteristic spectrum that only depends on their temperature (figure 12.5). A perfectly black object would naturally be in equilibrium with its radiation (i.e. at any wavelength it is emitting and absorbing the same amount of energy per second), so the spectrum has been named *black body radiation*[4]. (Many objects that look black to the eye are not perfectly black. Any object that looks black absorbs all visible light, but it may not absorb IR or UV radiation—a perfectly black object would absorb all wavelengths equally.)

As the temperature increases, so the wavelength at which the most electromagnetic radiation is produced decreases—the object glows red, then more yellow and finally white as the distribution of wavelengths produced falls within the visible spectrum.

Careful measurements have shown that Penzias and Wilson's radio radiation is equivalent to that which would be produced by a black body 'warmed' to 2.7 K. This radiation is now known as the cosmic background.

Since the initial discovery of the cosmic background was made, measurements of its spectrum have been refined both in accuracy and the range of wavelengths studied. In the early 1990s the highly successful Cosmic Background Explorer (COBE) satellite was able to measure the

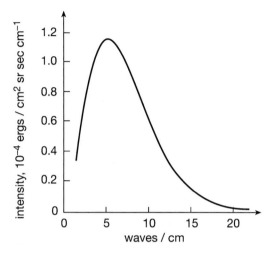

Figure 12.6. The results from the Far Infrared Absolute Spectrophotometer (FIRAS) instrument on the COBE satellite. The curve shows the expected intensity from a single temperature black body spectrum. The FIRAS data match the curve exactly, with uncertainties less than the width of the black body curve drawn. It is impossible to distinguish the data from the theoretical curve. (The COBE datasets were developed by the NASA Goddard Space Flight Center under the guidance of the COBE Science Working Group and were provided by the National Space Science Data Center (NSSDC) through the World Data Center A for Rockets and Satellites (WDC-A/R&S).)

radiation over a very wide range of wavelengths and showed that it followed the black body spectrum with remarkable precision.

12.2 Explaining the evidence

The modern theoretical science of cosmology started in 1915 with the publication of Einstein's theory of general relativity. In this paper Einstein expanded the range of his earlier special theory of relativity to include gravity. Shortly after he attempted to apply his new theory to the universe as a whole in the hope of showing that the equations produced a unique solution. Unfortunately they do not. In the early 1920s a rather obscure Russian meteorologist and mathematician Alexander Friedmann published a set of solutions to Einstein's equations that now form the basis of modern work on the evolution of the universe.

12.2.1 Redshift and the expansion of the universe

The Friedmann equations show that space is not an empty volume through which galaxies are moving—space *itself* is changing. If we observe the distance between two galaxies to be increasing in a Friedmann universe, it is not because the galaxies are moving apart; it is because the space is swelling[5].

Most people are not used to thinking of space as a *thing*. To try to understand this idea, imagine the surface of an inflated balloon that is covered with ants. The ants will move about on the surface—this is like galaxies moving *through* an unchanging space.

Now imagine that we put small piles of food on the surface of the balloon, so that the ants stay in place for an extended period. If we were to inflate the balloon further, the ants would all start to move apart as the surface of the balloon stretched. *But the ants would not be moving.*

It is the second case that is similar to Friedmann expansion—the ants are not moving *over* the surface, they are moving *with* it. The surface of the balloon represents the space in which galaxies are placed. Galaxies within the universe appear to be moving as the space in which they are contained stretches. Furthermore, as light passes through a stretching universe its wavelength will also be stretched. This is why the light from galaxies is shifted towards the red end of the spectrum.

Clearly it is very difficult to visualize the meaning of a phrase like 'space stretching'. One has to try to abandon the old idea that space is nothing more than an emptiness through which something moves. When one initially thinks about the universe expanding the natural reaction is to imagine the galaxies getting further apart as they move into areas of the universe that originally did not have galaxies in them (like a gas introduced into the corner of a box diffusing throughout the whole volume). This is *not* what is happening. I have tried to illustrate the difference in figures 12.7 and 12.8.

In figure 12.8 the space has been expanded, but not the galaxies themselves. The expansion of the universe would cause the galaxies, stars and us to expand as well if not for the local gravity and electromagnetism that holds us together. The expansion of space can

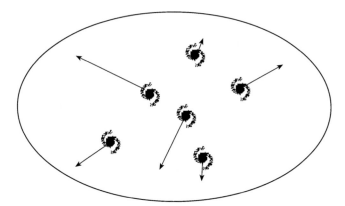

The universe as a box filling with galaxies

Figure 12.7. The natural idea is that galaxies are rushing apart filling previously empty regions of space.

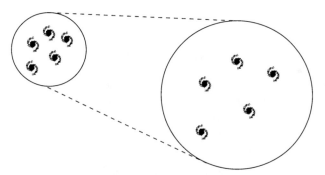

The universe as expanding space

Figure 12.8. The truth is that the universe is already full of galaxies. The space they occupy is getting bigger. In this diagram the bigger circle is a magnification of the smaller.

only be seen in the emptiness between galaxies and in the stretching of the light that crosses this region.

To see how this expansion gives rise to Hubble's law consider figure 12.9. In this diagram light is shown being emitted in our direction from two galaxies A and B. Each successive diagram is a larger

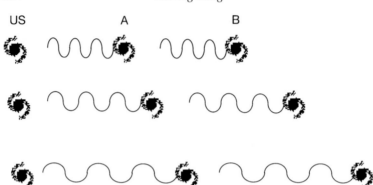

Figure 12.9. Hubble's law.

magnification of the first one. The wavelength of the light is increased by that magnification as well as the distance to each galaxy.

By the time that a 50% expansion has taken place, the light from galaxy A has arrived at us redshifted by 50%. However, the light from galaxy B is still on the way to us as it started off further away. Another 50% expansion will have taken place by the time that light from B arrives. Hence B will be considerably more redshifted than A because the universe has expanded by a greater amount in the time that it takes to get to us.

Consider also what is happening from the point of view of galaxy A. An observer based in galaxy A would consider themselves to be stationary and all the other galaxies to be moving away from them. Friedmann expansion implies that *all* galaxies will observe the others to be moving away from them—there is nothing special about our galaxy.

Hubble's law is a consequence of the expansion of the space through which light is travelling. However, it is often treated as an example of the Doppler effect, which is the shifting up in frequency of a sound produced by an object moving towards us, and down in frequency if the object is moving away. The same effect can be observed for any type of wave. It is possible to observe genuine Doppler effects in the light from stars, but the Hubble shift is not one of them.

Treating Hubble's law as if it was due to the Doppler effect is confusing as it lends support to the idea of galaxies moving *through* space—which we have been trying to avoid. However, it is useful to calculate the *apparent* speed at which galaxies are moving away from us. This tells us about the rate at which the universe is expanding.

We can calculate this apparent speed by combining Hubble's law with the Doppler relationship:

$$z = \frac{u}{c}$$

where u = speed of light source and c = speed of light. Hubble's law states:

$$z = H'd$$

or using the Hubble constant, H, the law can be written as

$$z = \frac{H}{c}d$$

so

$$\frac{u}{c} = \frac{H}{c}d$$

or

➡ $$u = Hd.$$

Hence the apparent speed of the galaxies is also proportional to their distance from us.

If the universe is currently expanding then logically there must have been a time when all the galaxies were together in one place. At some point all the stars would be right next to each other—in fact at this stage there would be no stars, just a super-hot plasma of hydrogen and helium gas. If we go back further, unwinding the expansion of the universe, then all the atoms must merge together. Protons and neutrons lose their individual nature and the universe becomes a sea of quarks, gluons and leptons. Particle physics becomes important. Further back in time still we start to go beyond what we currently understand. In principle the density of the universe continues to climb without limit as everything gets closer and closer together. This is the era of the big bang—the mysterious event that started the whole expansion going. The existence of this extraordinary starting point is a logical consequence of the observed expansion according to Friedmann's mathematics.

We can use the expansion law to estimate the time that has passed since the start of the expansion.

$$\text{Velocity} = H \times \text{distance}.$$

If the velocity is constant, then the time it has taken a galaxy to reach its current distance is

➡️ $$\text{time} = \frac{\text{distance}}{\text{velocity}} = \frac{\text{distance}}{H \times \text{distance}} = \frac{1}{H}.$$

As briefly discussed earlier, measuring H is no easy task. The best currently accepted value is

$$H = 65 \pm 10 \text{ km s}^{-1} \text{ (Mpc)}^{-1}$$

which corresponds roughly to a value of 20 km s^{-1} (million light years)$^{-1}$ i.e. that a galaxy 1 million light years away would be expanding away from us at 20 km s^{-1}. Hence, if the expansion has taken place at a constant rate the time the galaxy has taken to get to 1 million light years away is:

$$
\begin{aligned}
\text{time} &= \frac{1 \text{ million light years}}{20 \text{ km s}^{-1}} \\
&= \frac{9.43 \times 10^{18} \text{ km}}{20 \text{ km s}^{-1}} \\
&= 14.9 \text{ billion years.}
\end{aligned}
$$

Hence the often quoted value of the age of the universe as being 15 billion years.

12.2.2 Helium and the background radiation

Friedmann's work provides a convincing explanation for Hubble's law—an explanation that comes naturally out of general relativity. No particle physics has had to be inserted into the model in order to explain Hubble's results.

Working backwards from the idea of cosmic expansion leads to the conclusion that the universe must have gone through a stage in which its density and temperature was very much greater than that observed now. This is the natural stage on which particle physics can play a role. From this come equally natural and compelling explanations for the helium abundance and the cosmic background radiation. However, before we can explore these ideas in depth, it is helpful to pin down exactly what we mean by various ideas—such as the temperature of the universe.

The temperature of the universe

As we chart the history of the universe it is convenient to mark our progress in terms of temperature. The temperature of a gas is a measure of the average kinetic energy of the particles within it. We shall be dealing with particles that are moving at speeds near to that of light, so it is more appropriate to refer to their total energies. For a gas the temperature is related to kinetic energy by

➡ $$\langle KE \rangle = \tfrac{3}{2}kT$$

with T = absolute temperature in Kelvin and k = Boltzmann's constant $= 1.38 \times 10^{-23}$ J K^{-1}.

The numerical factor, in this case 3/2, varies depending on the type of particle involved, but it is generally ~ 1 so we shall be working on the assumption that in the universe:

➡ $$\text{average energy, } \langle E \rangle \sim kT.$$

Thermal equilibrium

Gases are composed of particles moving with a wide variety of speeds. We are able to define a temperature for the gas because the particles are colliding with one another (and the walls of the container) in such a way as to keep the average energy constant (if some particles collide and speed up, some others collide and slow down—so the average remains the same). The gas is in *thermodynamic equilibrium*.

If the temperature of the universe as a whole is to mean anything, then the particles within it must be in thermodynamic equilibrium. However, the particles are not simply bouncing off one another, as happens in a gas—these collisions are particle interactions dictated by the four fundamental forces.

Particle physics helps us to understand the processes that kept the universe in equilibrium.

Threshold temperatures

For each particle species there is a threshold temperature. Once the universe drops below that temperature the species is effectively removed from the universe.

As long as the temperature of the universe is above threshold, the following reactions can take place:

$$\text{particle} + \text{antiparticle} \rightarrow \gamma + \gamma \qquad (12.1)$$

$$\gamma + \gamma \rightarrow \text{particle} + \text{antiparticle}. \qquad (12.2)$$

These reactions are similar to the annihilations and materializations that we have been studying throughout this book. While these reactions are taking place the particle and its antiparticle will be major constituents of the universe. Obviously the photons involved have to have sufficient energy to provide the intrinsic energy of the particles, so that

$$2E\gamma = 2mc^2$$

with m being the mass of the particle. Once the energy of the photons in the universe becomes too small for the materialization process to work (remember that the photons are always being cooled by the expansion of the universe: as they are shifted towards the red end of the spectrum their energy decreases) there will be no materializations to replace particles converted to photons by the annihilations, and that species is wiped out.

This will happen once the photon temperature drops below:

$$T = \frac{\langle E \rangle}{k}$$
$$= \frac{E\gamma}{k}$$
$$= \frac{mc^2}{k}.$$

As an example, consider the presence of protons in the universe. The threshold temperature of the proton is:

$$T = \frac{mc^2}{k}$$
$$= 1.1 \times 10^{13} \text{ K}$$

and so once the universe had cooled below this temperature the protons and antiprotons annihilated each other and the proton became a very rare object, compared to photons. Table 12.1 lists the threshold temperatures of various particles.

Table 12.1. Particle threshold temperatures.

Particle	Mass (MeV/c^2)	Threshold temperature $(\times 10^9 \text{ K})$
p	938	10 888
π^+	140	1620
e^-	0.5	6
μ^-	106	1226

Radiation versus matter

As the universe expands the density of matter decreases (the same amount of matter occupies a larger volume). Cosmologists find it convenient to chart the expansion of the universe by the *scale parameter*, S. If the size of the universe could be measured at a defined point in time (today would be convenient) the universe's size at any other time could be calculated by multiplying every distance by the factor S for that time (if today was the reference, then for all times in the past $S < 1$).

The volume of the universe varies with the scale factor S^3, so the matter density should vary as $1/S^3$. The same is true of the energy density that this matter represents (as they are related by $E = mc^2$).

The photons in the universe also contribute to its energy density. However, as the universe expands the photons are redshifted—their wavelength increases with the scale factor. The energy of a photon depends inversely on its wavelength, so the energy must vary as $1/S$. This is coupled with the expansion of the universe's volume, so the photon's contribution to the energy density of the universe must vary as $1/S^4$.

In the earliest phases of the universe the temperature was higher than threshold for all particle species. Hence matter and radiation were both important in the universe. Furthermore all the particles were moving at speeds close to that of light, so in all cases[6]:

$$E = \text{KE} + mc^2$$

but

$$\text{KE} \gg mc^2$$

∴

$$E \sim \text{KE}.$$

Hence the matter behaved similarly to the radiation. This earliest phase of the universe is called the *radiation-dominated* epoch.

As the universe ages the parts of the energy density due to matter and radiation continue to drop but at different rates, so eventually the energy density of matter becomes greater than that of radiation. The universe is now in the *matter-dominated* epoch. This happened a few hundred thousand years after the big bang.

Detailed calculations using Friedmann's models show that the rate of expansion of the universe is different in the two epochs. One specific model was promoted in a joint paper by Einstein and de Sitter. It is known as the *flat universe* (for reasons to be explained in chapter 13) and has the simplest dependence of scale on time:

➤   Radiation-dominated epoch $S \sim t^{1/2}$

➤   Matter-dominated epoch $S \sim t^{2/3}$.

The early history of the universe

Now, at last, we are in a position to chart the early history of the universe and to see how the helium abundance and cosmic background radiation arose.

We will start our account of the history of the universe at 10^{-43} seconds after the big bang. At this time the temperature of the universe was greater than 10^{33} K and gravity was as strong as the other three fundamental forces. This is the epoch of ignorance—our understanding of physical law fails us.

As the universe expanded and cooled towards 10^{33} K gravity decoupled from the other forces and the grand unified theories (GUTs) mentioned in chapter 11 apply. There are many competing theories and the physics of this epoch is far from certain. In broad outline the universe is above the threshold temperatures of the X, X̄, Y and Ȳ particles that mediate the GUT force. Consequently the universe is populated with quarks, leptons and all the exchange particles. The numbers of quarks, antiquarks, leptons and antileptons are kept equal by the GUT interaction. However, as we reach the end of the GUT epoch the X and Y particles disappear from the universe. As they decay they leave behind a small excess of quarks over antiquarks. Detailed calculations suggest that as this epoch ends the universe should contain $1 + 10^9$ quarks for every 10^9 antiquarks[7].

We have now reached 10^{-35} seconds after the big bang.

From now until 10^{-12} seconds the universe's temperature is above threshold for the W and Z particles and so the electromagnetic and weak forces are equal in strength—the strong interaction has separated. Quarks and antiquarks are too close to one another in the universe for the strong force to bind them into hadrons. Any quarks that did temporarily bind together would be easily blasted apart again by collisions with the high energy photons present in the universe. This epoch is referred to as the *quark plasma*. At the end of this period the universe has expanded sufficiently for the quarks and antiquarks to bind into hadrons. In particular protons and neutrons form. The tiny excess of q over q̄ left over from the GUT epoch is now reflected in a tiny excess of p over p̄ and n over n̄.

At 10^{13} K and 7×10^{-7} seconds into history, the universe drops below the threshold for protons and neutrons. As a result these particles stop being major constituents of the universe. There is a general annihilation of p with p̄ and n with n̄, but the universe is left with a small excess of matter over antimatter. Before the annihilation the number of protons and neutrons were kept equal to those of the photons by the equilibrium reactions:

$$p + \bar{p} \rightarrow \gamma + \gamma \qquad (12.3)$$
$$\gamma + \gamma \rightarrow p + \bar{p} \qquad (12.4)$$
$$n + \bar{n} \rightarrow \gamma + \gamma \qquad (12.5)$$
$$\gamma + \gamma \rightarrow n + \bar{n}. \qquad (12.6)$$

After the annihilation we expect one proton to survive for every 10^9 photons in the universe. This is a number that can be measured today and is found to be comparable to the predictions from GUTs. The surviving protons and neutrons are still kept in thermal equilibrium with the rest of the universe via neutrino interactions:

$$\bar{\nu}_e + p \rightarrow n + e^+ \tag{12.7}$$

$$\nu_e + n \rightarrow p + e^- \tag{12.8}$$

$$e^- + p \rightarrow n + \nu_e \tag{12.9}$$

$$e^+ + n \rightarrow p + \bar{\nu}_e \tag{12.10}$$

which are recognizable as those that we met in chapter 2 concerning the solar neutrino problem. At this early stage of the universe the density of matter and the energies concerned are big enough for neutrino interactions to happen regularly. This keeps protons and neutrons in thermal equilibrium and also the number of protons equal to the number of neutrons. As the neutrino energy drops the mass difference between the proton and neutron becomes significant and reactions (12.7) and (12.9) happen less often than the others. Therefore there is a gradual conversion of neutrons into protons as the universe ages.

At about 10^{-5} seconds the universe drops below the threshold for pions and muons and these particles cease to be major constituents of the universe.

At 1.09 seconds after the big bang the temperature has dropped to 10^{10} K and the density is such that the neutrinos can no longer interact with matter. From this instant on they decouple from the universe. They do not disappear. They remain forever within the universe but do not interact with matter to any extent. The expansion of the universe redshifts their energy just as it does for photons. They retain a black body energy spectrum and should presently have a temperature of approximately 2 K. Although we expect there to be ~500 neutrinos per cm^3 in the universe, we currently have no way of detecting them. Measuring the temperature of the neutrino background would be an important experimental confirmation of the big bang.

Once the neutrinos have decoupled, there is no process to keep the protons and neutrons in thermal equilibrium—reactions (12.7)–(12.10) stop. By this time the ratio of protons to neutrons (p/n) has shifted to

82% protons and 18% neutrons. From this time onwards, the decay of free neutrons will start to eat into their numbers further.

The next major milestone takes place 3 seconds after the big bang. This is the time at which the universe has cooled to 5.9×10^9 K and consequently the e^- and e^+ threshold has been reached. There is a net annihilation of e^+ and e^- leaving an excess of e^- due to the unequal numbers set up in the GUT epoch. The remaining e^- stay in thermal equilibrium with the photons as they are free charges and hence interact well with electromagnetic radiation.

3.2 minutes into the life of the universe the photon temperature drops to the point at which deuterium (pn nuclei) can form. Before this time any deuterium nuclei that formed would be blasted apart by photons. We have entered the epoch of nucleosynthesis in which nuclear reactions such as the ones below take place.

$$p + n \rightarrow d + \gamma \tag{12.11}$$
$$d + d \rightarrow {}^3He + n \tag{12.12}$$
$$d + d \rightarrow {}^3H + p \tag{12.13}$$
$${}^3He + d \rightarrow {}^4He + p \tag{12.14}$$
$${}^3H + d \rightarrow {}^4He + n. \tag{12.15}$$

The primary result of this 'cooking' is the production of helium-4, but in addition some deuterium is left over and some heavier elements, such as beryllium and lithium, are made by a series of reactions and decays.

During nucleosynthesis all the free neutrons are swept up and bound into nuclei. Once bound in this way, the strong interaction between the protons and neutrons stabilizes the neutrons preventing them from decaying. By the time that nucleosynthesis starts, the decay of free neutrons has shifted the p/n ratio to 87/13. Out of every 200 particles 26 neutrons will combine with 26 protons to form 13 helium nuclei. This leaves 148 protons, so the mass ratio of the nuclei produced is:

$$\frac{13 \times 4}{200} = 26\%$$

in very good agreement with the observed fraction of helium in the universe. This is one of the most important experimental confirmations of the big bang theory.

More refined calculations show that the exact proportions of the elements produced depend on the density of protons and neutrons and the expansion rate of the universe during the period of nucleosynthesis (both influence the rate at which particles can interact with each other). Helium comes out in the region of 24–25%, deuterium about 0.01% and lithium 10^{-7}%.

Nucleosynthesis stops when the temperature has dropped below that required for nuclear reactions, which happens 13 minutes into history.

Nothing much happens now for the next 300 000 years or so. The universe continues to expand and cool until, coincidentally, two events happen at more or less the same time.

Until this time the energy of photons in the universe has been sufficient to ionize any hydrogen atoms that form from protons and electrons. Now the photon energy drops below this value and the electrons rapidly combine with protons. When this happens the universe stops containing large numbers of free electrical charges. There is nothing left for the photons to interact with and they decouple from matter. The universe is said to have become *transparent*. This event is called the *recombination* of electrons (although strictly speaking as time flows forwards from the big bang the electrons have never combined with the nuclei before!). These photons do not disappear. Prevented from interacting with matter to any great extent they continue to exist in the universe and are steadily redshifted by the expansion. Some 15 billion years later, they are detected by a microwave antenna and become known as the background radiation.

At about the same time, the universe also stops being dominated by the density of radiation and becomes matter dominated.

This then is the universe that we live in now—at least according to the traditional cosmological theories outlined in this chapter. The big bang idea has proven to be tremendously successful in explaining the three pieces of evidence that ushered in the modern era of observational cosmology, but it does leave some questions unanswered. Aside from the philosophical issues that always arise when discussing the creation of the universe, there are some practical matters as well—such as how the various structures (stars, galaxies etc) arise. One might at first assume this to be a technical matter not closely related to particle physics, but as we shall see over the course of the next few chapters, the large-scale structure of the universe, its future and its origins may be deeply related to the theory of forces and particles in the microworld.

12.3 Summary of chapter 12

- Hubble discovered that the light from distant galaxies was shifted towards the red end of the spectrum and that the size of the shift was proportional to the distance to the galaxy;
- the observed amount of helium in the universe is to great for it to have been manufactured inside stars;
- there is a uniform microwave background radiation characterized by a black body spectrum of approximately 3 K;
- the universe is in a state of expansion in which the space between galaxies is being stretched;
- light waves crossing this space will also be stretched—explaining Hubble's law;
- if the universe is expanding it follows that at one time it was much smaller than it is now;
- at earlier times the universe was so dense and hot that particle interactions took place commonly;
- the temperature of the universe can be characterized by the energies of the particles in it;
- various processes keep the particles in thermal equilibrium above critical temperatures;
- below its critical temperature a particle stops being a major constituent of the universe;
- above the critical temperature the particle and its antiparticle are kept in thermal equilibrium;
- GUT theories suggest that as the critical temperature is crossed there may be a small excess of matter over antimatter formed;
- helium is manufactured in the early stages of the universe in just the right amount to agree with the current measurements;
- after hydrogen atoms form the photons of the universe decouple and cool due to the expansion to become the 3 K background.

Notes

[1] You are probably aware that each atom has its own pattern of wavelengths that it emits, e.g. sodium strongly emits a pair of wavelengths in the yellow part of the spectrum. However, a solid block of sodium when warmed will glow red like any other material. This is because the atoms of sodium are bonded together which has the effect of turning the line spectrum of the individual atoms into a continuous emission over a range of wavelengths. The same sort of thing is happening inside stars.

[2] Joseph von Fraunhofer (1787–1826).

[3] It is a little known fact that the elements of which we are made were first built out of hydrogen and helium inside a star that exploded billions of years ago scattering its debris across the galaxy. It is from this debris that our planet and our bodies have been formed.

[4] Of course, in the case of the cosmic microwave radiation we are talking about an equilibrium between the universe of matter and radiation in the earliest epochs—there is nowhere outside the universe to emit and absorb radiation from!

[5] Galaxies have two motions—their proper motion *through* space and their *Hubble flow* due to the universe's expansion. In strict terms the proper motion is the only movement of the galaxy, the Hubble flow is due to the expansion of space between the galaxies.

[6] In order to avoid confusion between kinetic energy and temperature I have used KE in these equations instead of the more normal T.

[7] I did say that it was a *small* excess.

Chapter 13

The geometry of space

Much of our current understanding of the early universe comes from the application of Einstein's general theory of relativity. The general theory of relativity is widely acknowledged as one of the most beautiful creations of the human mind. It is also one of the most fiendishly difficult physical theories to master. An adequate treatment, even at the level aimed for here, would require a book of its own[1]. The best that we can manage at the moment is a brief overview concentrating on the main ideas that are needed for cosmology.

13.1 General relativity and gravity

13.1.1 When is a force not a force?

The most radical aspect of general relativity is its removal of the idea that gravity is a force. Right now I am sitting on a rather comfortable chair in front of my iMac computer (grape coloured) typing this chapter and I am acutely conscious of the force of gravity on me. Reaching out my hand to pick up a pen I have just accidentally pushed a book onto the floor. The book's fall provides another seemingly incontrovertible piece of evidence for the existence of a gravitational force.

Of course no one is denying the everyday occurrences and experiences of the sort just referred to; what is at issue is the interpretation of them. Consider the book falling to the floor. A casual examination of the situation would suggest that the book was initially stationary and then was pushed sideways by the force exerted on it by my arm. Once the book had slid free of the desk's surface the force of gravity was able to pull on the book without the opposing force of the bench acting upwards to prevent it falling. Free to move vertically the book now falls to the floor.

This is the sort of analysis of the situation that is routinely taught at schools and that most people with a vaguely scientific education would be willing to accept. However, we have already seen that the everyday interpretations can lead to problems when pushed too far. The interaction force between my arm and the book when analysed in sufficient detail would lead to the exchange of off mass shell photons between charges in my arm and the book. Physics is a game that is played at various levels. What works well for playing with books falls far short when it comes to dealing with rotating neutron stars!

One of the intuitions that led Einstein to his theory relates to a simple matter of gravitational law—all things fall under gravity at the same rate. Supposedly this was demonstrated first by Galileo when he dropped two similarly shaped, but different mass objects (say a melon and a cannon ball) off the side of the leaning tower of Pisa. They were observed to strike the ground at the same time. Being the same basic shape and size the objects could be assumed to have the same force of air resistance acting on them, so it was a fair test. When the air resistance can be neglected, or circumvented in this way, it is clear that all objects irrespective of their mass reach the ground together when dropped from the same height.

When considered from the standpoint of Newton's laws of gravity and motion this is a subtle but simple point. Einstein, however, saw a greater significance—to him it implied that the motion was not simply the same for every object that was falling, it was independent of the object. In a sense the path is built in to the space the objects are moving through— like rails on which a train is moving. Which object happens to be currently riding those rails is consequently irrelevant.

Einstein replaced the idea of force with the idea of geometry. To him the space through which objects move has an inherent shape to it and the objects are just travelling along the straightest lines that are possible given this shape. In order to follow this idea through it is important to have some tools for describing the shape of space.

13.2 Geometry

The properties of space independent of the objects embedded in it are characterized by a branch of mathematics known as geometry. The most ancient master text of geometry is that written by Euclid in which he explored the basic properties of points, lines and the geometrical figures (squares, triangles etc) that could be constructed from them. His work was based on a series of simple assumptions the first five of which are listed below:

1. A straight line can be drawn between any two points.
2. Any straight line that stops at a particular point can be extended to infinity.
3. A circle of any given radius can be drawn round any given point.
4. All right angles are equal.
5. Given an infinitely long line L and a point P, which is not on the line, there is only one infinitely long line that can be drawn through P that does not cross L at any other point (this is clearly the line that is *parallel* to L).

And so it goes on. These assumptions define what it is to do geometry in the sense that Euclid meant it. Using these assumptions and logically rigorous arguments Euclid went on to prove a variety of statements about geometrical figures—such as the following.

- The angles in a triangle when added together must add up to 180°.
- The circumference of a circle divided by its diameter is a fixed number called π.
- In a right angled triangle the lengths of the sides are related by $c^2 = a^2 + b^2$ where c is the length of the side opposite to the right angle.

The third of these statements is the well known theorem of Pythagoras. Euclidean geometry is a beautiful exercise in formal structure and logic—but it was also intended to be something more. There is no doubt that Euclid thought that he was also saying something demonstrably true about the nature of the space in which we live. His geometry was describing the universe as well as being a formal mathematical system. The theorems that he proved helped to describe the nature of space. It would be impossible in Euclidean space to draw (accurately!) a triangle with an angular sum that was not 180°.

Euclidean geometry as it was first conceived is a geometry of two-dimensional objects, but it is quite easily extended into three dimensions. Geometrical figures now include planes and spheres with their own sets of theorems that describe them (for example the surface area of a sphere is $4\pi r^2$), but they always conform to the basic rules set out in the assumptions.

13.2.1 Breaking the rules

Mathematicians can be a curious breed and eventually someone decided to see what would happen if the assumptions of Euclid were not all taken as read. This is an interesting exercise for mathematicians who like taking pieces out of formal systems to see what breaks. The obvious starting point, to a mathematician, is statement 5 as this is the most 'contrived' sounding one in the list. It turns out that there are two ways of changing statement 5 without destroying the whole system[2]:

5L Given a line L and a point P not in the line, there are *at least two* other lines that can be drawn through P that do not meet L at some other point. This gives rise to a class of geometry called Lobachevskian after the mathematician Lobachevsky.

5R Given a line L and a point P there are *no* lines through P that do not meet L at some other point. This leads to Riemannian geometry (after Riemann).

As formal mathematical exercises one can go on to prove various mathematical theorems on the basis of the Euclidean list with just statement 5 replaced. Some re-interpretation is required, but that is not a difficult exercise (see box).

A similar exercise can be carried out to show what Lobachevskian geometry looks like in two dimensions.

Just as Euclidean geometry can be extended into three dimensions, so can the alternatives. However, here we come against a serious problem of imagination and illustration. A three-dimensional Euclidean space needs no real illustration as it corresponds to the space of our everyday experience. However, in order to draw a three-dimensional space that was Riemannian one would have to have access to a four-dimensional world in which the three-dimensional surface of a four-sphere could be drawn. A very common mistake is to think that the three-dimensional

An example of non-Euclidean geometry in two dimensions—the surface of a sphere

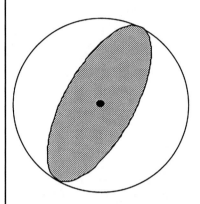

In two-dimensional Riemannian geometry a line is part of an arc on the surface of a sphere. To ensure that the line is the shortest distance between two points, the arc must be part of a great circle—i.e. a circle like the one shown shaded that cuts through the surface of the sphere and its centre. The line is part of the arc where the circle cuts the surface of the sphere.

Here we have an example of how Euclid's fifth statement is different on the surface of a sphere. Given the line L (which must be part of a great circle to count as a line in Riemannian geometry) and another point P which is not on the line, it is impossible to drawn a complete line (i.e. one that comes back to P again) that does not cut through L at some point.

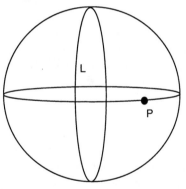

Riemannian surface is just that shown in the box above. The geometry dictated by the Riemannian rules is the geometry of the *surface* of the sphere—*which is two dimensional curved in three dimensions*. In order to show that we have had to draw a sphere as it would look in a three-dimensional space (which is Euclidean). What we are talking about now is a whole three-dimensional space (curved in four dimensions!).

Einstein's theory replaces gravity as a force with the notion that space can have a different geometry from the Euclidean. Just as a line drawn on the surface of a sphere seems curved when viewed from the vantage point of a three-dimensional space (but it is actually the straightest line that can be drawn given the geometry of the surface), so an object moving through space is following the straightest line that it can. If we had access to 'higher dimensions' then we could see this just as easily.

Surprisingly it is easier to follow the implications of this for the universe than it is to see how it applies to the book falling on the floor. So our first task will be to see what can be said about the geometry of our universe as a whole.

13.3   The geometry of the universe

One of the problems of dealing with the general theory of relativity is the difficulty of coming up with a clear physical picture of what is happening. This is especially hard when one is trying to consider the universe as a whole. Fortunately, the basic features of a particular geometry translate from the three-dimensional case to two dimensions. We have done this before when considering the expansion of space. In that case we used the idea of ants crawling over the surface of a balloon to help. Our picture of ants can be helpful here as well, provided a few salient points are appreciated.

Our universe is, of course, three dimensional. The ants in our picture are crawling over a two-dimensional surface[3] (the surface of the balloon). We can observe their behaviour and appreciate that the balloon is getting bigger quite easily as we are looking at their two-dimensional world from the vantage point of our three-dimensional one. From the ant's point of view, things are rather different. The third dimension is not part of their universe—they have no access to it. They cannot move off the surface of the balloon, neither can they move inside to its centre. They are constrained only to live on the surface. The only way they can notice the expansion of their universe is by the fact that objects in it are getting further apart. This is exactly the same way that we appreciate the expansion of our universe. We cannot access a 'fourth dimension' from which we can see the expansion of our three-dimensional world.

I have used this picture of ants on the surface of a balloon often in teaching about the big bang and I have grown used to having to explain it quite carefully. Questions that nearly always come up are:

1. Is the centre of the balloon the place where the big bang happened (is it the centre of the universe)?
2. If the universe is expanding, then it must be expanding into something—is that not the three-dimensional space surrounding the surface of the balloon?

Reasonable answers to these questions would be along the following lines:

1. The centre of the balloon is not part of the ant's universe. By definition they cannot get to it—it has no connection with any part of their universe. That means that the big bang did not take place at that point. The big bang in the ant's universe happened everywhere on the surface at the same time. The same is true in our universe—there is no centre, the big bang happened everywhere at once.
2. One can quite happily imagine that the ant universe is expanding into the third dimension that we inhabit, but that is of no real help to the ants. The volume of three-space in which their universe is embedded is not part of their universe and they cannot move into it. The only space they know is the surface of the balloon and that surface is stretching. In the same way, we can imagine our three-dimensional universe is embedded in a four-dimensional hyperspace, but it does not help us in any way. This embedding space is not part of our universe, and there would seem to be no way of gaining access to it[4].

Models and analogies like the anty one that we have been considering are the lifeblood of scientists who are trying to create visual pictures of the physics that is being related to them by their mathematics. General relativity provides equations which can be solved to show how the universe in which we live evolves with time. The solutions describe universes in which the three-dimensional space is curved. This is a perfectly well defined mathematical concept, which we can visualize in two dimensions (like the surface of the ant's balloon) but we are not equipped to imagine what a three-dimensional universe would be like curving in some embedding four-space. Such things are not part of our everyday experience, so the languages that we use are not equipped to describe them. We are forced to describe them in terms of more visualizable things, which are not always totally appropriate.

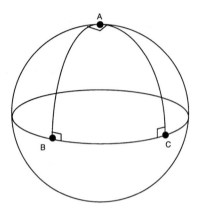

Figure 13.1. A triangle on the surface of a sphere can contain three 90° angles! This is a classic example of the difference between Riemannian geometry and Euclidean.

A universe that closes back on itself, like the surface of a balloon does, is called a *spatially closed* universe and it corresponds to the Riemannian geometry mentioned in the previous section.

The ants could demonstrate that their universe is spatially closed by starting to crawl off in a given direction. After some time, if their universe is closed (as we know it to be) the intrepid ant explorers would arrive back exactly at their starting point. In principle, if our three-dimensional universe were spatially closed, we could send a spacecraft out in a given direction and, after an unfeasibly long journey, it would arrive back at where it started. If this turned out to be the case, then the three-dimensional universe would be curved round in four dimensions in a closed surface.

A journey round the universe could never really be contemplated, so it is fortunate that there are other ways of determining the geometry of the space in which we live. At school we learn that the angles of a triangle if added together will sum to 180°—at least they do in Euclidean geometry. On the surface of our balloon, this is not the case!

Figure 13.1 shows how this simple rule of geometry depends crucially on the curvature of the surface on which the triangle is drawn. Triangle

ABC starts at the 'pole' of the sphere and has part of the equatorial line as one of its sides. This means that the triangle contains three right angles and so has an angular sum of 270°. In principle this could be checked in our universe using a large enough 'triangle'—perhaps by shining laser beams between space probes. Experiments of this sort are being planned for the future.

Now, so far we have been discussing spatially closed universes as if that were the only option. However, Friedmann's analysis of the solutions to Einstein's equations suggest that there are three different possibilities. The spatially closed one we have now covered in some detail—think of it as being the universe in which a crawling ant would eventually return to its starting point. On the other hand, Friedmann universes can also be *spatially open*—an ant crawling over a two-dimensional spatially open universe would never return to its starting point, but it would never reach an 'edge' either for such a universe has to be infinite in size.

Spatially open, infinite universes come in two different varieties depending on their curvature. Such a universe can be *flat*—in two-dimensional terms an infinite sheet of flat paper—and contain triangles that always have angle sums of 180°. Such a universe has a Euclidean geometry. This was not a possibility covered by Friedmann, but a flat universe model was championed by Einstein and de Sitter in a paper they wrote a few years later.

The other possibility is for an open universe to be curved, but not to close back on itself. Writers have struggled for a long time to describe such a shape in words, and the best description is to compare the two-dimensional version to the curved leather of the sort of saddle used when riding a horse.

A curved open universe of this sort would contain triangles with an angular sum that was less than 180° as specified by the Lobachevskian geometry.

In summary then, the intrinsic 'geometry' of a universe can be appreciated by any beings that happen to live inside the universe. One does not need to have access to any 'higher' dimensions within which the curvature of the space is clear to see. Simple geometrical experiments (testing the angles of a suitably large triangle) will show the inhabitants

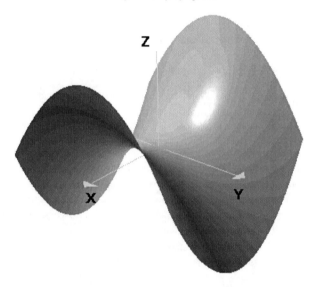

Figure 13.2. A portion of a two-dimensional open universe that is curved. The shape is often compared to that of a saddle used by horse riders. X, Y and Z are coordinates of the three-dimensional space in which the two-dimensional surface is embedded. It is the surface that represents a curved universe.

what sort of universe they live in. Visualizing the curvature of a two-dimensional universe would be as impossible for the flat ants living in it as comprehending that our three-dimensional universe has curvature is to us. Flat ants would have to rely on the advice of expert ant mathematicians just as we have to appreciate the calculations of our cosmologists in order to give some meaning to the observations.

13.4   The nature of gravity

It is all very well talking about the geometry of a universe, but we have still not tackled the basic problem of how the fall of a book can be described in terms of geometry. Once again two-dimensional illustrations are a key help in understanding what is happening.

A ball rolling along a flat (Euclidean) two-dimensional surface would roll (without friction) in the same direction forever at a constant speed.

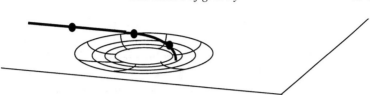

Figure 13.3. A ball rolling along a flat surface encounters a dint in that surface and is deflected so that it rolls into the dint.

However, the presence of a dint in that surface would deflect the path of the ball as it passed.

Now, in reality it is gravity acting that pulls the ball down the curvature of the surface and the ball is itself a three-dimensional object so we have to be a little careful about the analogy. However, it is possible to make the connection between the distortion of the flat geometry that the ball was rolling along and the new path it has taken. From the point of view of general relativity, a two-dimensional point sliding along the plane should replace the ball. The fact that the point then follows a curve in the vicinity of the dint is simply due to it having no choice—it cannot leave the plane in which it is sliding any more than we can leave the three-dimensional world in which we live. One must also ignore the fact that the dint in the sheet forces the point to travel vertically downwards. Once again, up and down have no meaning in the two-dimensional world that the point inhabits. A flat ant observing the point's antics would not comment on any movement in the third dimension. Remember that any light rays that allow the ant to see the ball would also be constrained to travel along the (curved) surface of the sheet as well!

According to general relativity the space surrounding a lump of matter acquires a geometry that is different from Euclidean and which determines the paths that objects follow near to this mass[5]. In this view a spacecraft being deflected from its path as it passes a large mass (planet) is simply following the straightest line that it can through a curved space.

Returning to the example of the book falling, we now have a picture of a book on the table that is sharing the path of a planet orbiting a star. The book, along with the planet, is following a path through a curved volume of space that is determined by the geometry of the situation. However the book is also in a more local region of space, which is curved by the mass of the planet nearest to it (like a dint inside a dint). The ball is trying to follow the altered path due to the planetary distortion, but is prevented from doing so by the (genuine) force of the table acting on it. However, once released it will follow a path that is slightly different from that which the whole planet is proceeding along, which results in it falling to the floor (from our point of view).

But why does it speed up?

It is easier to relate how general relativity determines the geometry of the universe, as time does not come into it. By a lucky turn of physics the geometry of space for the universe as a whole can be described without having to include time as well. For the motion of the ball this is not the case.

General relativity is actually a theory of space and time. We have already seen how Euclidean geometry can be extended from the plane to three dimensions quite easily; it could even be taken into four dimensions. Einstein's theory includes time as a fourth dimension. This is not the same as the four-dimensional hyperspace we were considering earlier when describing the curvature of space. What we are now considering is a four-dimensional spacetime which is also curved (and to draw that we would need a five-dimensional hyperspace!). One of the 'directions' along this surface represents time. It is a very radical thought, but time can be regarded on a similar footing to the other three dimensions and included into the geometry dictated by general relativity[6]. Indeed this has some observable effects—clocks run more slowly close to large masses than they do when further away. This is the direct physical effect of the curvature of time.

A ball falling to the ground is moving through a region of space that is spatially fairly Euclidean (the curvature of space really comes into effect with much stronger gravitational fields), but which has a notable curvature of time. The ball is following the straightest path that it can through a curved spacetime geometry—something that determines its path and its speed.

13.5 The future of the universe?

According to conventional big bang cosmology, the energy density of the universe determines its future. Friedmann's equations predict that if the energy density of the universe is greater than a critical value, the expansion will eventually stop and then reverse itself.

It has become conventional to express the energy density of the universe as a fraction of this critical energy density. Ω is defined as the ratio between the measured density of the universe and the critical density. While it is difficult to measure the actual density of the universe, the critical density required to reverse the expansion can be rather easily calculated[7].

If $\Omega > 1$ then the universe contains enough mass for its energy density to be greater than the critical value. If our universe turns out to be like this, then it is spatially closed (like the surface of the balloon). Its future will be like that of a balloon universe that expands to a maximum radius (but remember the radius is not part of the universe) and then contracts again. While the universe is expanding the light from distant galaxies is redshifted by the stretching of the space through which it is passing. As the universe starts to contract again, the light will be blueshifted. Eventually the universe would change back to conditions similar to those just after the big bang—extreme temperature and pressure with elementary particles interacting. This has been termed the *big crunch*. A universe with $\Omega > 1$ is both spatially closed (in principle you can walk right round it) and also *temporally closed* (it will not last for ever).

On the other hand, if $\Omega < 1$ then the universe will continue to expand for ever. The universe is spatially and temporally open. The galaxies continue to sail away from each other as the space between them stretches. The expansion rate slows as time passes, but it gets closer and closer to a finite non-zero value depending on the density of the universe.

The specific value $\Omega = 1$ fixes the universe on a unique path that expands for ever, but at a rate that steadily approaches zero—cosmologists term this a *flat universe* (this is the Einstein–de Sitter model mentioned before). Such a universe is infinite in extent and duration with a Euclidean geometry.

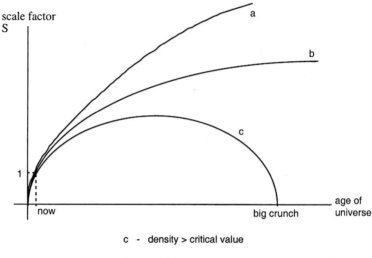

c - density > critical value

b - density = critical value

a - density < critical value

Figure 13.4. Possible futures of a universe following a simple Friedmann model. *Do not take this seriously as a graph displaying the ultimate fate of OUR universe.* Recent results to be discussed in chapter 15 suggest that the universe is rather more complicated than the simple Friedmann models would suggest.

Clearly it is of paramount importance to measure the energy density of the universe. This is not an easy task (to put it mildly). There are several ways of approaching this problem.

One good way is to measure quite precisely the amount of primordial (i.e. formed during the big bang) helium, deuterium, lithium and beryllium present in the universe. The abundances of these elements are very accurately predicted from the physics of nucleosynthesis during the big bang and depend on the amount of matter present in the early universe (cosmologists often refer to this as an estimate of the amount of *baryonic matter*). A careful study of the spectra of old stars and nebulae has pinned the baryonic density, as a fraction of the critical density, to being $\Omega_B = 0.043 \pm 0.003$.

Another way of estimating the baryonic mass of the universe is to estimate the total mass of stars that it contains. Such measurements suggest that the total amount of luminous matter in the universe is actually about ten times less than the baryonic matter estimated from primordial elements. Consequently it would seem that most of the baryonic matter in the universe is not in the form of stars; most of it is in the form of large clouds of gas (mostly hydrogen of course).

An estimate of the density of the universe can be obtained by looking at the gravitational forces acting within galaxies and within clusters of galaxies. Such measurements will be discussed further in the next chapter; however, their results suggest that the total density of matter in the universe places Ω rather nearer to 0.3!

Cosmologists and astronomers have had to come to terms with this rather surprising result. If all this matter were baryonic, then the primordial abundances would be utterly different to those observed. So, this mass cannot be in the form of baryons—it is called *dark matter*[8]. At present there is no conclusive theory about what this material is. Several possibilities are discussed in chapter 14.

Yet another estimate of Ω can be obtained from very precise measurements of Hubble's law. If the universe is slowing down, or speeding up, then the line drawn through the data should stop being straight for the most distant objects. Studying this enables us to pin down the density compared to the critical density. Recent supernova studies strongly suggest that $\Omega = 1$. This is an extraordinary result, which has been confirmed by detailed inspection of the cosmic microwave background. If Ω is 1, which seems now almost certain, then the baryonic matter that we have spent the last 12 chapters studying is less than 5% of the total density of the universe.

13.6 Summary

- General relativity describes gravity as a curvature of space and time rather than a conventional force;
- the curvature of space is best described by a branch of mathematics known as geometry;
- there are three different types of geometry that are relevant to the study of the universe as a whole;

- Euclidean geometry is the geometry of a flat universe which is, in two-dimensional terms, a flat piece of paper with an infinite extent;
- a closed universe has the same geometry as that of the surface of a balloon (Riemannian);
- open universes have a surface that is curved like the shape of a saddle, in two dimensions (Lobachevskian);
- when objects move through a curved space they travel along lines that are the straightest paths possible through that space;
- general relativity also describes the curvature of time—which is another way of saying how clocks vary in speed from place to place. This helps determine the motion of the objects both in terms of its direction and speed;
- the geometry of a universe depends on its energy density;
- a universe with an energy density equal to the critical value is a flat universe and will expand indefinitely, although at a rate that tends towards zero;
- open universes will expand forever, but their expansion tends towards a finite value that is non-zero. They are also spatially infinite;
- closed universes are doomed to re-collapse eventually and have a density greater than the critical value. They are finite in extent, but unbounded;
- modern estimates place the universe short of the critical density, but indicate that there is a large amount of dark matter present.

Notes

[1] Now there's an idea!

[2] In mathematical terms what this means is that these statements can be substituted for statement 5 in Euclid's list and a sensible system that does not contain any contradictions can still be assembled.

[3] For the purposes of this picture, we have to imagine a species of flat ant—in other words a two-dimensional ant that slides around on the two-dimensional surface of the balloon. A genuine three-dimensional ant would be poking out of the model universe!

[4] Neither is it mathematically necessary. All the properties of a space can be fully described by using a system of reference that lies inside the space. For

example, we manage to specify positions on Earth quite easily using a system of latitude and longitude. One could also do this by specifying angles from the centre of the planet—this would be like using an embedding space system.

[5] This is often illustrated by imagining that the plane is a rubber sheet. Placing a large mass on the sheet causes a distortion similar to the dint in figure 13.3. I regard this as being unhelpful as it raises all sorts of problems with the mass being three-dimensional and the plane being two-dimensional as well as the slightly awkward point that it is gravity pulling the mass down that causes the dint that is being used as an analogy for gravity!

[6] Strictly speaking what is represented is *ct*—the speed of light multiplied by time—this gives a 'length' that can be represented geometrically.

[7] The critical density at any period in the universe's history is given by $\rho_c = \frac{3H^2}{8\pi G}$ where H is the value of the Hubble constant at that period and G is the universal gravitational constant.

[8] We may not know what it is, but we do know a fair amount about its properties. Clearly it can exert a gravitational pull (or we would not be able to detect it). However, it does not interact with electromagnetic radiation—the gravitational estimates tell us that galaxies have about ten times more dark matter than the conventional stuff in them. If dark matter did interact with light, it would be a struggle to see the galaxies at all!

Chapter 14

Dark matter

In this chapter we shall survey the evidence for the existence of
dark matter in our universe and see how it plays a crucial role
in our understanding of how galaxies and clusters of galaxies
are formed. This in turn will open up other issues that need
resolving in the next chapter.

14.1 The baryonic matter in the universe

In the previous chapter I indicated how important it is to have some
estimation of the density of the universe, at least compared to the critical
value. For some years now astronomers have been building evidence that
the amount of visible matter falls a long way short of the critical density.

Looking for elements produced in the big bang, as distinct from being
made inside stars since then, has turned out to be a very sensitive way of
estimating the amount of baryonic matter present in the early universe
(and presumably it is all still here!). Detailed calculations relating to
the nuclear reactions going on during the era of nucleosynthesis have
shown that the abundances are very critically dependent on the baryonic
density. As we have seen, the primary result of these reactions is the
production of hydrogen and helium, but several other elements were also
produced—although in tiny fractions compared to the big two.

Estimating the proportions of these various elements is not an easy task
and many factors have to be taken into account.

- Helium-4—picking up primordial helium-4 is tricky as it is also made inside stars. The best measurements come from regions of hot ionized gas in other galaxies from which the ratio of hydrogen to helium can be deduced. These gas clouds are bolstered when old stars explode, scattering elements into the galaxy. Consequently they contain both primordial elements and other isotopes made since then. However, the amount of oxygen present in the gas cloud (which is not made primordially) serves to indicate how much the cloud has been fed by exploding stars. Knowing this enables astronomers to extract the amount of helium-4 that must also have come from the stars, leaving an estimate of the primordial stuff. The best estimates put the helium abundance at $24.4 \pm 0.2\%$ of the universe by mass.
- Deuterium is an especially good isotope to look for. If the density of protons and neutrons in the early universe is large, then there will be a faster reaction rate producing deuterium. This happens for two reasons. Firstly there are more protons and neutrons to react. Secondly as the density of the universe is comparatively high, the expansion rate will be slower and so the particles are closer together promoting the reaction as well. Small changes in the proton and neutron density make quite large changes in the amount of deuterium produced. Not only that, but as there is no known mechanism for producing deuterium in stars, or the space between stars, it is rather easier to estimate the primordial abundance. The best recent estimates of primordial deuterium come from the study of the spectra of very distant objects called *quasars*[1]. These measurements have helped pin down the baryon density as a fraction of the critical density to being $\Omega_B = 0.043 \pm 0.003$ (for a Hubble constant value of 65 km s^{-1} (Mpc)$^{-1}$). The policy now is to use the deuterium measurements to establish the baryon density and from that to *predict* the density of the other elements which are then checked against observation.
- Lithum-7—Some stars destroy lithium, while others produce it. The best values for the primordial lithium abundance come from a study of the oldest stars in our own galaxy. The various reaction pathways that can produce lithium in big bang nucleosynthesis produce two baryon densities consistent with the currently estimated abundance.

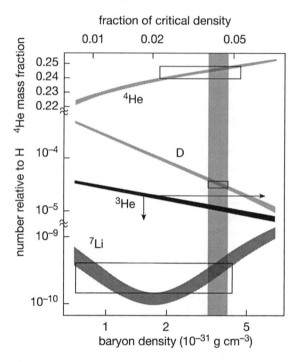

Figure 14.1. Nucleosynthesis in the big bang produces various light elements with abundances that depend on the Hubble constant (here fixed at 65 km s^{-1} (Mpc)$^{-1}$) and the density of baryonic matter (protons and neutrons) in the early universe.

Figure 14.1 summarizes the currently accepted values for the light elements. Also shown are the predictions of how the abundances vary for various baryonic matter densities (all for a Hubble constant of 65 km s^{-1} (Mpc)$^{-1}$). The vertical band shows the value of Ω_B consistent with all these figures, that being 0.043 ± 0.003.

14.2 The evidence for dark matter

Astronomers have a wide variety of different tools available to them to study the universe. They have optical telescopes on the ground and in orbit round the Earth. Radio telescopes are able to examine

Keyhole Nebula

Hubble Heritage

NASA and The Hubble Heritage Team (STScI) • Hubble Space Telescope WFPC2 • STScI-PRC00-06

Figure 14.2. This spectacular Hubble Space Telescope image demonstrates how much of our conventional information comes to us electromagnetically. This is a picture of a nebula (a cloud of gas) in our own galaxy about 8000 light years away and 200 light years across. The large roughly circular feature contains bright filaments of hot, fluorescing gas (giving off electromagnetic radiation), and dark silhouetted clouds of cold molecules and dust (blocking out electromagnetic radiation from behind them). (These data were collected by the Hubble Heritage Team and Nolan R Walborn (STScI), Rodolfo H Barba' (La Plata Observatory, Argentina), and Adeline Caulet (France).)

objects that cannot be seen optically as well as giving a different perspective on the visible objects. The same is true of the γ-ray, x-ray, infrared and ultraviolet satellites that have been crucial in developing our understanding. However, all these instruments rely on electromagnetic radiation. Radio waves, microwaves, infrared radiation, visible light, ultraviolet light, x-rays and γ-rays differ only in their wavelength; they are all the same type of radiation. A type of matter that did not interact with or produce any form of electromagnetic radiation would not be seen at all.

This is the issue that astronomers and cosmologists now face. The amount of evidence available for the existence of *dark matter* (in its most general terms matter that cannot be directly detected electromagnetically) has become compelling. Not only does it seem that this stuff exists, it also appears to be the dominant matter component of the universe.

There are at least four ways of inferring the presence of dark matter—from the gravitational effect that it has on the motions of stars in galaxies, the motion of galaxies within clusters, gravitational lensing and the effect that it has on the expansion of the universe.

14.2.1   The motion of stars in galaxies

The stars in the arms of a spiral galaxy (see figure 14.3) are rotating about the centre of the galaxy. They are held in orbit by the gravity of other stars in the central galactic bulge.

Astronomers can study the rate of this rotation by looking at the Doppler shift of the light coming from various parts of the galaxy[2]. As the stars move towards us their light is Doppler shifted towards the blue part of the spectrum. As they move away, the light is redshifted. By comparing light from one edge of the galaxy with that from the centre and the other edge of the galaxy, astronomers can establish the rotation curve (a graph showing the speed at which stars are travelling against distance from the centre of the galaxy) for the stars within it. Figure 14.4 shows the rotation curve for a typical spiral galaxy.

The first obvious feature of the curve is the steady increase in rotation speed up to about 10 kpc.[3] There is nothing unexpected in this. The central bulge of this galaxy has a radius of about 10 kpc so the stars up to this distance lie inside that region. They are only influenced by any stars that are nearer to the centre of the galaxy than they are. Clearly as you move out from the centre there are more and more stars behind you, so the overall gravitational pull on you actually increases with distance inside the region of the bulge. Consequently the more slowly moving stars tend to drift inwards towards the centre of the galaxy. As the galaxy evolves stars settle into place within the bulge and tend to be moving faster the further out they are[4].

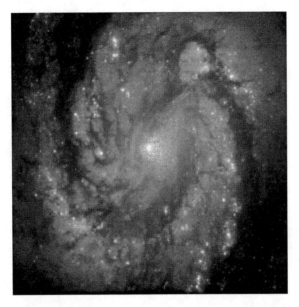

Figure 14.3. A typical spiral galaxy (courtesy Hubble Space Telescope Science Institute). The spiral arms and the densely populated central bulge are clearly visible.

Once the central bulge is passed however, the situation should change. The outer arms of the galaxy are far less populated with stars than the central bulge (where most of the galaxy's mass is concentrated). As one moves further out the mass pulling back on you is hardly changed by stars that are nearer than you are (as there are comparatively few of them compared to those in the central bulge). A much bigger effect now is the decrease of the gravitational pull with distance. Consequently one would expect the speeds of the stars to decrease with distance from the centre outside the bulge (there is not enough gravity to hold on to the fast ones so they tend to drift outwards slowing down as they go[5]).

A small drop in speed beyond 10 kpc can be seen on the graph, but it does not continue to fall as we might expect—it stabilizes to a virtually constant value. This might be an indication that there was more mass present at the edges of the galaxy. However, a simple visual inspection shows that this is not true—unless the mass is invisible.

Dark matter

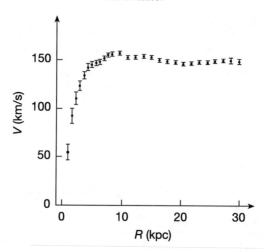

Figure 14.4. The rotation curve for the galaxy NGC3198 (a spiral galaxy that can be found in the constellation of Ursa Major). From Begeman (1989).

Aside from the problem inherent in the flattening of the curve, the stars are also generally moving too quickly. Computer simulations of galaxies with stars moving at typical measured speeds show that the galaxy is unstable and quickly starts to break up. There is not enough visible mass to hold it together at such high speeds.

Both effects can be explained if the galaxy is surrounded by an enormous halo of matter that extends for many times the visible radius of the galaxy. As we do not see such a halo in our telescopes it must be of very dark matter (i.e. it does not interact with or produce much light of its own).

It has been estimated that the amount of dark matter needed in a typical halo is between two and four times the amount of visible matter seen in the galaxy[6]. The part of the galaxy that we can see represents only a quarter of its total structure, in mass terms.

14.2.2 The motion of galaxies in clusters

Astronomers can also look at the redshift of light from galaxies within a cluster and measure the speed of the whole galaxy relative to its partners.

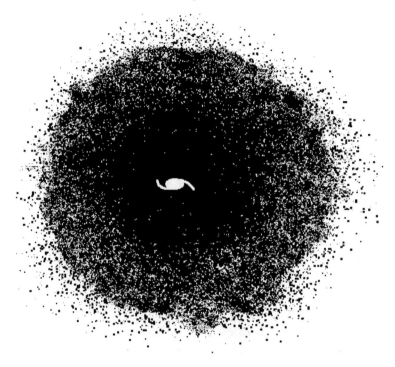

Figure 14.5. A halo of dark matter surrounds a galaxy holding it together by gravity.

Our own galaxy is a member of a small cluster that includes the nearest galaxy to ours Andromeda (2.2 million light years away), the large and small Magellanic Clouds (galaxies visible in the southern hemisphere and first observed by the explorer Magellan) as well as something like 20 other small galaxies. Measurements indicate that Andromeda is moving towards our galaxy with a speed of 100 km s^{-1}.

This is an uncomfortably large speed. At this sort of speed Andromeda would have long ago left the local cluster if not for the combined gravitational pull of all the other members. Most of this pull is coming from the Milky Way (its nearest partner). To explain Andromeda's presence we have to assume that the halo of dark matter surrounding our galaxy extends millions of light years beyond the edge of the visible galaxy (or about 70 times the visible radius) and *contains a mass of dark matter equal to ten times the mass of visible matter.*

The story does not stop here. Our local group of galaxies is apparently being pulled towards a large supercluster in the constellation of Virgo[7]. This is an enormous grouping of thousands of galaxies about 60 million light years away. Measurements suggest that the Milky Way is moving towards the Virgo supercluster at a speed of 600 km s^{-1}. This indicates that the total mass of the supercluster could be up to 30 times that of the visible matter.

It seems that every time we increase the scale of our observations we see evidence for dark matter. Not only that but the proportion of dark matter increases with the scale. This is not too upsetting, as it is only the amount of matter between objects that affects their motion. However, it does suggest that each galaxy may only be a tiny pinpoint within a truly enormous volume of dark matter—possibly stretching tens of millions of light years (over a thousand times the visible radius).

14.2.3 Gravitational lensing

The bending of the path of light rays by massive objects was one of the earliest predictions that Einstein produced from his new theory of general relativity. The idea was used in a very early confirmation of the theory. Einstein realized that light coming to us from a distant star would have to skim past the Sun for part of the year. Six months later the Earth would be on the other side of its orbit and so we would have a clear line of sight to the star without the Sun in the way. By comparing the position of the star relative to others on two occasions, six months apart, one could measure the extent to which the light was deflected when it had to pass the Sun. Needless to say, Einstein calculated this deflection and was not surprised when it was shown to be correct[8].

Einstein also predicted that large masses in deep space could distort the light from other objects and act in a similar manner to an optical lens. At the time, Einstein did not hold out much hope for ever observing such effects, but he failed to anticipate the dramatic improvements brought about in the design of optical telescopes (especially having the Hubble Space Telescope in Earth orbit) as well as the advent of radio astronomy. We now have numerous examples of gravitational lenses.

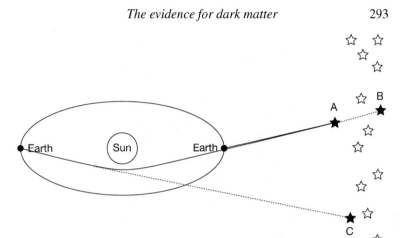

Figure 14.6. The bending of light by the Sun. For part of the year the position of star A in the sky seems to be close to that of star B. However, six months later the light from A is deflected on its way past the Sun, making A now appear to be near star C.

If the lensing object, the light source and the Earth are in near perfect alignment, then the light from the source is distorted into a ring image round the lens (see figure 14.7). Such perfect alignment is, of course, very rare but several spectacular examples have been found. More often the alignment of the lens and light source is not exact. When this happens a multiple image is formed rather than a ring (see figure 14.8).

Gravitational lenses can also be formed when collections of massive objects provide the gravitational effect, rather than one compact massive object. Clusters of galaxies can lens the light from more distant clusters. When this happens the combination of multiple masses can lead to complex visual effects including segments of arcs and multiple imaging (see figure 14.9). Working backwards from the image astronomers can deduce a lot of information about the mass in the lensing cluster and how it is arranged. Such studies are complicated, but the general trend in the discoveries is always that the cluster contains more mass than can be accounted for by the visible component.

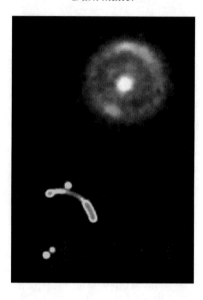

Figure 14.7. Two images of the same object. The top image in infrared light shows a bright central galaxy, which has another more distant galaxy behind it. The lensing effect of the galaxy's mass has bent the light of the more distant object so that a near perfect ring image is formed round the lensing galaxy. The bottom image is the same object in radio wavelengths. The radio image ring is not complete showing that the alignment of the objects is not as perfect as it seems in infrared. The top image is from the Hubble space telescope and the bottom image from the UK's Merlin radio array (reproduced with permission of the Merlin VLBI national facility, University of Manchester).

14.2.4   The large-scale structure of the universe

The Friedmann models assume that matter is distributed smoothly throughout the whole universe. This is an excellent approximation in the early moments of history, and appears to hold when we examine the distribution of clusters of galaxies at the scale of the whole universe. However, when we look at the universe at smaller scales we find that it is rather lumpy—there are planets, stars, galaxies and clusters of galaxies. The universe is not uniform at every scale. It has immense structures within it.

Figure 14.8. In this example of a gravitational lens a bright distant object (a quasar) has been lensed by a nearer elliptical galaxy (visible as the faint haze round the central cross). The alignment of object lens and Earth is not perfect, so a multiple image (the cross) has been formed rather than a complete ring. The number of images formed depends on the shape and mass distribution of the lensing object. (J Rhoads, S Malholva, I Dell'Antonio, National Optical Astronomy Observatories/National Science Foundation. Copyright WIYN Consortium Inc., all rights reserved. Used with permission.)

When astronomers turned their attention to the details of galaxy distribution in the universe they found that they congregated into clusters. At a higher level of structure clusters merge together into superclusters. These are huge regions of the universe in which the number of galaxies per unit volume is far higher than the average. Often these superclusters are strung out in long filaments.

Between these superclusters are gigantic volumes known as voids that contain virtually no galaxies. Typically these voids can be 150 million light years across and contain ten times less than the average number of galaxies.

Figure 14.10 shows the latest results of a survey of galaxy distribution in a slice of the sky. The 2df team aims to map 250 000 galaxies by the time that survey is completed[9]. Each galaxy has its position in the sky recorded and its distance away from us is deduced from the redshift of its light. The map produced shows two slices cut into the universe as seen from an observer placed at the apex of the two wedges and contains galaxies out to 2.5 billion light years from us.

Gravitational Lens in Abell 2218   HST · WFPC2
PF95-14 · ST ScI OPO · April 5, 1995 · W. Couch (UNSW), NASA

Figure 14.9. Gravitational lensing by a galaxy cluster (courtesy Hubble Space
Telescope Science Institute). The light from a group of distant (and hence
older) galaxies has been distorted by passing through the gravitational field of
a nearer cluster of galaxies. The result is a series of arcs and multiple images
that can reveal information about the distribution of matter in the nearer cluster.
(Courtesy W Couch (University of New South Wales), R Ellis (University of
Cambridge) and NASA.)

Clearly visible in this map are the large-scale filaments and voids
representing structure on the largest scales. As yet we cannot explain
fully why these structures have formed. It seems clear that matter
clumped together on a small scale to form galaxies that were then drawn
together into clusters, but the detailed processes at work are still being
worked out.

There is a great deal of information about the state of the universe to
be deduced from maps of this kind. Some aspects of this we shall pick
up a little later when we focus again on the growing evidence for Ω
being equal to 1. However, one thing that is of immediate relevance is
the analysis that the 2df team have done on the gravity at work within
clusters.

The general redshift of the galaxies arises from the Hubble expansion
of the universe. Correspondingly distant galaxies should have a greater

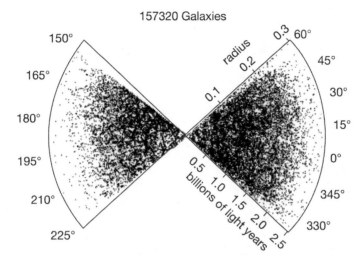

Figure 14.10. A recent map of a slice through the universe (February 2001). The observer is placed at the point where the two wedges meet. Each dot on this map represents a galaxy. The density of observations falls off towards the edges, as the most distant galaxies are faint and hard to observe. (Courtesy 2df galaxy redshift survey team.)

redshift, according to the general pattern established by Hubble's law. However, if the redshifts are compared for galaxies on the edges of clusters then deviations from this general trend can be seen. Galaxies on the far side of a cluster have a redshift that is slightly less than the average redshift of galaxies in the cluster (which is established by the general Hubble flow). Such galaxies are being pulled back by the gravity of the matter in the cluster. On the other hand, a galaxy on the near side of the cluster is slightly more redshifted than the average as it is being accelerated away from us and into the cluster. Careful analysis of the data produces an estimation of the cluster's mass. Once again clear evidence for the existence of dark matter is found.

14.2.5 Conclusions regarding dark matter

In summary, the evidence for dark matter is quite compelling. This evidence is seen in a variety of different ways and at various scales—from the stars in individual galaxies to the large-scale structure of the universe as a whole. As we moved from studying galaxies to clusters the proportion of dark matter present increased. However at the largest scales the evidence stabilizes implying that Ω_{matter} is 0.3 in total. Now,

$$\Omega_{matter} = \Omega_B + \Omega_{otherstuff}.$$

Ω_B can be further broken down into $\Omega_{visible}$ and $\Omega_{darkbaryonicmatter}$ (as we shall see shortly dark baryonic matter can be clouds of hydrogen, small objects like the planet Jupiter etc). Ω_B has been accurately measured by the abundances of various elements produced in the big bang nucleosynthesis and is taken to be ~ 0.04. $\Omega_{visible}$ is estimated to be less than 0.01. In total then, the dark matter of the universe (which is comprised of some baryonic matter as well as more exotic stuff that has yet to be defined), seems to be about 30 times the amount of visible matter in the universe. This is a quite stunning conclusion.

The dark matter in the universe seems to have two components to it. Certainly some of it must be in the form of baryonic matter that, for some reason or other, is not clearly visible in the universe. However, the nucleosynthesis data show that there cannot be anything like enough of this stuff to bring Ω up to 0.3. The rest must be non-baryonic material of some form. There has been quite an amount of speculation as to what this might be—the most likely idea being some form of particle dredged up from the slightly more exotic theories such as supersymmetry. The crucial aspect to this that is missing is direct observational data on particles of dark matter. At the present time, no such evidence exists. There are many experiments going on currently trying to detect dark matter, but as yet without success. It is important then that the information from cosmology and astronomy be milked as much as possible to try to pin down the properties of this mysterious stuff.

14.3   What is the dark matter?

In order to proceed further in understanding the nature of the dark matter, some other constraint needs to be applied.

At this point it pays to pick up another thread in this entangled story.

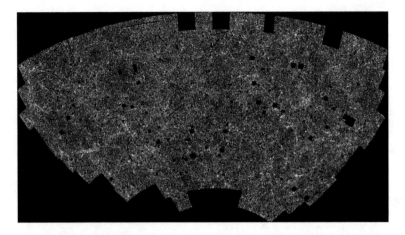

Figure 14.11. The APM galaxy survey picture. This image contains 2 million galaxies as seen from Earth. The slice represents about one-tenth of the visible sky as seen from the south pole. The gaps in the picture are where data had to be eliminated due to the presence of a bright object in our own galaxy getting in the way. Large-scale structuring can clearly be seen. The APM galaxy survey is a computer-generated sky survey of over 2 million galaxies and 10 million stars, covering about one-tenth of the whole sky, in the South Galactic Cap. It is derived from scans of 185 photographic plates taken with the UK Schmidt Telescope at Siding Spring, Australia. These plates were scanned with the Cambridge APM (Automated Plate Measuring) laser scanner. (Courtesy Steve Maddox, Will Sutherland, George Efstathiou, Jon Loveday and Gavin Dalton.)

14.3.1 The problem with galaxies

The 2df galaxy map shows that the universe is filled with structure at the largest possible scale. Figure 14.11 is another illustration on the same lines.

This image is a slice across the sky (and so contains no depth information) but clearly shows the clustering of galaxies into filaments and the presence of voids between clusters. Astronomers and cosmologists wish to understand how these sorts of features can evolve in the universe and it turns out that dark matter has a crucial role to play.

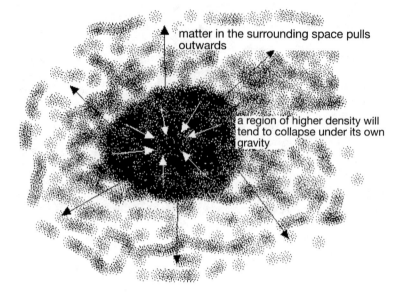

matter in the surrounding space pulls outwards

a region of higher density will tend to collapse under its own gravity

Figure 14.12. A region of slightly higher than normal density will expand more slowly than the surrounding space.

The only way for these structures to form is if the matter in the early universe was not quite as smooth as assumed in the Friedmann models. In the very early universe any regions of space that had a slightly higher density than average would expand slightly more slowly than the rest (held back by the gravity within them). The density of matter in the space surrounding such a region would decrease relative to that within the region as the universe expanded. As long as the matter density surrounding the region stays greater than half of the density within the region, the gravity of the surrounding matter is pulling outwards on the matter inside sufficiently to stop it collapsing under its own gravity. However, eventually the outside drops to less than half the inside density and at this point the matter inside the region starts to collapse in on itself. As the matter collapses its density increases. This causes the gravitational effect to grow. The region is now in a state of catastrophic collapse and will form galaxies and the stars within them.

A great deal of recent effort has been spent trying to find evidence for these small density fluctuations in the early universe. Evidence for them should be imprinted on the cosmic background radiation.

Until the epoch of recombination (about 300 000 years after the big bang) electromagnetic radiation was kept in thermal equilibrium with matter, as there were plenty of free charges about to interact with the radiation. After recombination there were no more free charges and the photons decoupled from the matter. Since then they have been steadily redshifted and now form the cosmic background radiation.

In chapter 12 we saw that this radiation could be given an equivalent temperature. However, there is no reason to expect that this temperature will be exactly the same in every part of the sky. In fact there are good reasons to expect small differences caused by variations in the density of matter present at the time of recombination. If these temperature variations can be accurately mapped, they should give us useful evidence about the density variations in the early universe that eventually gave rise to galaxy formation[10].

After various attempts using ground-based instruments, the breakthrough discovery was made using the COBE satellite. The results it obtained show that there are fluctuations in the temperature of the background radiation between different parts of the sky. The satellite was able to measure the temperature of the sky in patches about 7° across and compare the temperatures of different patches[11]. The fluctuations measured are very small—*no bigger than a hundred thousandth of a degree*. Furthermore, the regions in which the temperature seems constant are huge—about 100 million light years across. The largest structures seen on the earlier sky maps would comfortably fit inside the blobs on this image.

Over the past few years other experiments have come along (notably the *Boomerang, Max* and *Maxima* balloon experiments) that have refined the discoveries made by COBE. They have been able to compare much smaller patches of sky than COBE could and so have seen the background radiation in much greater detail (but, unlike COBE, they have not been able to map the whole sky). The importance of these more recent results will be taken up in the next chapter. For the moment, our main concern is not with the areas of those regions on the sky that show

COBE–DMR Map of CMB Anisotropy

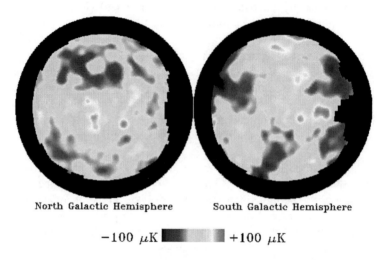

North Galactic Hemisphere   **South Galactic Hemisphere**

$-100 \ \mu K$ ▬▬▬▬ ▮ $+100 \ \mu K$

Figure 14.13. The COBE sky map. This image is derived from an analysis of the data obtained over the 4 year duration of the mission. Unlike early images that appeared in the press, the fluctuations shown here are an accurate visual rendition of the temperature variations as they would appear on the sky. (The COBE datasets were developed by the NASA Goddard Space Flight Center under the guidance of the COBE Science Working Group and were provided by the National Space Science Data Center (NSSDC) through the World Data Center A for Rockets and Satellites (WDC-A/R&S).)

variations—for galaxy formation the problem lies with the size of the temperature variation.

As we have seen, a region will start to collapse if the average density of the matter it contains is twice that of the matter surrounding it. Assuming that galaxy formation is happening right now, there must be parts of our current universe that show this 2:1 density variation. Working back from this, cosmologists can estimate the density ratio that this would have implied at the time of recombination. The ratio of the density in a region to that in the outside space *increases* by the same factor as that by which the universe's temperature *decreases* over the same period.

Minimum density variation at recombination

We can write the ratio of the inside density to the outside density as:

$$\frac{\text{density inside region}}{\text{density in surrounding space}} = 1 + x.$$

Now, as the universe expands the value of x increases in proportion to the change in temperature.

$$\text{temperature at recombination} = 3000 \text{ K}$$

$$\text{temperature now} \qquad = \qquad 3 \text{ K}$$

$\therefore x$ is now 1000 times bigger than it was at recombination.

The higher density region will start to collapse if the ratio of the inside density to the surrounding space $= 2$

$\therefore$ in order for collapse to start x must equal 1.

If regions are starting to form galaxies now, $x = 1$ in these regions implying that at recombination $x = 1/1000$

$\therefore$ the variations in density that we should see by looking at the COBE results should be at least $1/1000$ or galaxies could not have formed by now.

For galaxy formation to be going on now triggered by gravitational instability, there must have been regions of the universe at the time of recombination that were of a greater density than their surroundings by one part in a thousand. One would expect to see this reflected in the temperature fluctuations in the COBE sky map. Unfortunately the biggest fluctuations that have been seen in the background radiation are of the order of a hundred thousandth of a degree!

The situation is actually much worse than this. We have just calculated the *minimum* density variation required to *start* galaxy formation now. Yet when we look into space we see quite a lot of galaxies already formed! These must have been triggered by much larger density fluctuations at recombination, which grew to the collapse point only a few hundred thousand years after recombination. This places them way outside the limits set by COBE.

14.3.2 Dark matter to the rescue

The ripples in the cosmic microwave background indicate that the distribution of matter in the universe at recombination was too smooth to trigger the formation of galaxies. Dark matter provides the basis for a solution to this problem.

Non-baryonic dark matter may have started to form large clumps of higher than average density some time before recombination. We would see no evidence for this in the COBE data. Remember that, by definition, dark matter has to be of a form that does not interact with electromagnetic radiation. This means that vast clumps of the stuff could have formed before recombination without leaving any imprint on the background radiation. Such clumps of dark matter would pull gravitationally on the ordinary matter. However, the pressure of radiation in the universe would prevent the ordinary matter from falling into the dark matter clumps until after recombination. Once the radiation and the matter stopped interacting, the ordinary matter would be free and would then be mopped up by the dark matter clumps triggering the gravitational collapse that leads to galaxy formation. This could happen quite rapidly after recombination.

In rough outline this is how cosmologists believe that the galaxies and clusters of galaxies evolved from the imprints seen in the COBE data. For this to work it is essential that the dark matter started to clump together under its own gravitational attraction before recombination took place.

To make any further headway with a theory one has to identify what the dark matter is, as the manner in which it clumps together depends on its nature.

14.3.3 Types of dark matter

Two distinct classes of dark matter have been considered—*hot dark matter* and *cold dark matter*. In this instance the terms do not refer to temperature in the conventional sense. Hot dark matter is assumed to have been moving at close to the speed of light when the clumping process started. Cold dark matter, on the other hand, was moving at speeds very much slower than light as it started to form clumps.

There is no conclusive evidence that can directly show what the dark matter in the universe is, so the various candidates have to be examined theoretically with the aid of computer models. The purpose of these pieces of software is to simulate part of the early universe. Various mixes of dark and ordinary matter are programmed in and the simulated universe is allowed to expand. As the programme proceeds galaxies start to form from the clumped dark matter. Astronomers are looking for the mix that gives the most realistic simulation of the superclusters and voids that we see in the real universe.

In the early 1980s hot dark matter was all the rage. Neutrinos were studied in the role of hot dark matter. Calculations had shown that there should be about 100 neutrinos per cubic centimetre left over from the big bang. Even if neutrinos had a very small mass, in total they would contribute an enormous mass to the universe. Furthermore they only interact via the weak force. Being neutral they cannot interact with electromagnetic radiation. They satisfy all the requirements for dark matter.

Just at the time that cosmologists were in need of a form of dark matter, the particle physicists found massive neutrinos that would do the job. The results of experiments to measure the mass of the electron-neutrino were suggesting that they had a mass in the region of 30 eV/c^2. These experiments subsequently proved impossible to reproduce and the question of neutrino mass is still open. At the time, however, they were another piece of persuasive evidence. An industry grew up overnight as simulated universes populated with massive neutrinos were put through their paces.

Unfortunately, they did not work out. With such a small mass in each neutrino they would still be moving at essentially the speed of light. This means that only very large clumps of neutrinos have enough mass to hold together gravitationally. In a universe dominated by hot dark neutrinos the density variations would be regions of mass equivalent to a hundred trillion stars (about a thousand galaxies). Any smaller regions would rapidly be smoothed out by the fast moving neutrinos[12]. Not only that, but the pressure of these neutrinos would interact with the ordinary baryonic matter to tend to smooth out any density irregularities there as well.

With regions of such size collapsing under their own gravity the largest structures form first—the superclusters. This is when the theory starts to go wrong—it takes too long for the neutrinos to clump together on smaller scales to form galaxies within the clusters (the initial cluster would be a huge cloud of gas). Computer simulations show that superclusters should not start to form until about 2 billion years after the big bang, and then galaxies would only condense out of those structures comparatively recently. This does not square with the experimental evidence. Not only have there been plenty of quasars observed with redshifts so great that the light must have been travelling towards us for 10/11ths of the universe's age (i.e. they must have been formed less than 2 billion years after the big bang), but the Hubble Space Telescope's Deep Field image (figure 14.14) shows a population of very ancient galaxies—perhaps formed only a few hundred million years after the big bang.

In a hot dark matter universe, galaxies should only be forming about now. So, after a brief period of success and hope, interest in hot dark matter started to fade as physicists realized that they could not explain galaxies in this way[13].

Inevitably interest turned to cold dark matter. This slow moving stuff can clump on much smaller scales—dwarf galaxies would form and gravitationally pull together and merge to form larger galaxies, which would then gather into clusters. The theory is much more promising than hot dark matter.

Unfortunately, there is no single particle that is an obvious candidate for the role of the dominant constituent of the cold dark matter. Theorists have come up with a variety of exotic objects that have the general features required. The collective name coined for the various candidates is WIMPS (weakly interacting massive particles). The requirements are that the particle should not be able to interact electromagnetically, it has to be very massive (and hence slow moving) and be generated in sufficient quantities in the big bang. There is quite a list of candidates.

Hubble Deep Field
Hubble Space Telescope · WFPC2

PRC96-01a · ST ScI OPO · January 15, 1995 · R. Williams (ST ScI), NASA

Figure 14.14. The famous Hubble Deep Field image. The picture was assembled from 276 separate frames taken over a ten-day period in late 1995. As a result, some of the faintest objects ever seen show up. This portion of the full image represents a fragment of the sky about 1/120th of the size of the full Moon. Remarkably this tiny portion of the sky contains thousands of galaxies. Further analysis has revealed several dozen objects that may be incredibly distant (and therefore old) galaxies—a handful seem to be older than the most distant quasars. Information like this can be used to rule out the hot dark matter scenario of galaxy formation, and places severe constraints on any other theory. (Courtesy Robert Williams and the Hubble Deep Field Team (STScI) and NASA.)

Axions

Axion particles were suggested in the late 1970s as a way of tidying up some aspects of QCD. If they exist, axions should have an incredibly small mass, of the order of 10^{-5} eV, but they would have been produced in copious quantities in the big bang. Furthermore, the processes leading to their production would ensure that they were nearly stationary (even though they are very light in mass).

Photino

The supersymmetric partner of the photon. None of the supersymmetric particles have their masses fixed by the theory, but the fact that they have not been observed yet suggests that the lightest of them (thought to be the photino) must have a mass of at least 15 GeV.

Quark nuggets

The ordinary hadrons that make up the visible matter in the universe are composed of u and d quarks. The s quark (and the others) are too massive and quickly decay. However, it has been suggested that a nugget of matter produced in the big bang with roughly equal numbers of u, d and s quarks might just be stable. The key here is the size. Quark nuggets would be composed simply of quarks packed close together and could be between 0.01 and 10 cm in size. This would give them a mass between 10^3 and 10^{15} kg!

Baryonic matter

Dark matter is normally taken to be some exotic form of elementary matter, but in the widest sense some of the dark matter that we infer from its gravitational effects could be in the form of ordinary baryonic matter that has evaded direct electromagnetic detection. In the terminology of cosmology baryonic matter is any ordinary form of matter made from protons and neutrons. The visible matter that we can see in the universe—stars, dust clouds, nebulae etc—is all baryonic matter. It is possible that there is more of this stuff in the universe than we realized.

As we have seen the amount of baryonic matter has been fixed by the proportions of primordial elements produced in the big bang. These measurements are widely accepted in the astronomical community, but they still leave some room for more material than we can directly see. The limit placed by the primordial element production is $\Omega_B = 0.04$: estimates of the visible baryonic matter are somewhat less at $\Omega_{visible} < 0.01$. This leaves quite a lot of scope for dark forms of baryonic matter. There are several possible candidates for the baryonic contribution to dark matter.

One possibility is that galaxies are surrounded by clouds of hydrogen that are not big enough to block out the view, but there are enough of them to have a gravitational influence.

Another idea is that that there are more *brown dwarf* stars than we had previously thought. Brown dwarfs are starlike objects that were not large enough to heat up sufficiently to light nuclear reactions. They are failed stars. The search for these objects has recently intensified with the discovery of several microlensing events. When a massive compact object, such as a brown dwarf, passes in front of a background star, the gravity of the object can bend the light round it and produce a brighter image of the star while the object is passing in front of it. The gravity involved is not strong enough to produce distortion or multiple images, but a detectable change in the apparent brightness of the star can be measured (e.g. figure 14.15).

Brown dwarf stars are not the only things capable of microlensing in this manner—there are also white and brown dwarfs, neutron stars and black holes. Collectively these objects are known as MACHOS (massive astrophysical compact halo objects—as they may exist in the visible halos surrounding galaxies).

White dwarfs are the burnt out embers of old stars. Neutron stars are slightly more exotic objects that can be formed at the end of a star's life if it explodes. Black holes (as virtually everyone must know by now) are the remains of very massive stars that have collapsed under their own gravity. The gravitational forces of these objects are so great that not even light can escape from them. They may be enormously massive, yet smaller than Earth. Stephen Hawking suggested that mini-black holes (smaller than atoms) could have been created in the big bang. It is possible that they could be contributing to the dark matter. However, Hawking also showed that such mini-black holes should eventually

Dark matter

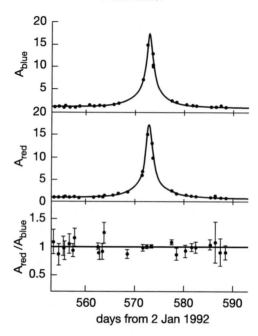

Figure 14.15. The light curve from a star in our galaxy being microlensed by another object. The graphs show the brightness of the star over a period of time in red and blue light (the ratio of the two also confirms that this is something affecting all the components of the star light in the same manner). The data in these graphs have come from the MACHO project to look for MAssive Compact Halo Objects, which may form a significant fraction of the dark matter in the Milky Way's halo. (Courtesy the MACHO science team. Further information can be found on their web page listed in appendix 4.)

explode in a gigantic burst of γ-rays. No evidence for this has been seen, so it is doubtful that they exist.

As measurements are refined so the consensus opinion of the extent to which baryonic matter of some dark form is contributing to the galactic halo has varied. Currently the general opinion is that there are not enough of the brown dwarfs etc to account for the mass necessary to explain the rotation curves.

This is not the total list of candidate cold dark matter objects, but it does show the sort of ingenuity that is being applied to the problem.

Cold dark matter started to come into vogue in 1982. The computer simulations rapidly showed that it was highly successful at explaining galaxy formation. Galaxy-sized structures would form before a few billion years of computer universe time had passed with some objects starting to appear after only a few hundred million years. By 1983 cold dark matter had become the accepted way of explaining galaxy formation, a position it holds to the present time.

Using cold dark matter models it is now possible to roughly outline the processes that lead to galaxy formation.

14.3.4 Making galaxies

Cold dark matter (of whatever variety) and ordinary baryonic matter produced in the early universe were distributed highly uniformly throughout the space, but small density variations were present from place to place. Ten thousand years or so into history the Friedmann expansion had increased the size of these variations and the dark matter started to clump gravitationally[14]. Ordinary matter was prevented from collapsing as the electromagnetic radiation is interacting with it and 'blows away' any matter clumps that start to form. However, at about 300 000 years into history recombination took place and the matter was freed from this effect. The COBE results reflect the variations in the matter density present at this time, but there are unseen dark matter variations that are much bigger. The ordinary matter was then pulled by gravity to these dark matter clumps. When the ordinary matter gathered onto the cold dark matter clumps it tended to collect in galaxy-size masses. In a volume of space containing both types of matter, ordinary matter soon separates from cold dark matter. The ordinary matter radiated energy away electromagnetically (triggered by collisions between atoms)—and slowed down. Gravity then pulled this matter into the centre of the cloud where it condensed further to form stars and galaxies.

What you end up with, as confirmed by computer simulations, is a large clump of cold dark matter with a glowing galaxy in the middle. This is exactly what we see in the evidence for dark matter—the galactic halos.

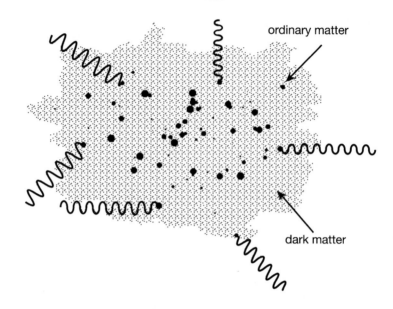

Figure 14.16. Ordinary matter can condense inside dark matter by radiating energy away.

Furthermore the gravitational interactions between neighbouring proto-galaxies that are forming at the same time, tend to set them spinning in opposite directions[15]. The amount of spin that can be imparted in this way, however, is not enough to prevent a forming galaxy from collapsing into a blob—spirals would not form. However, as the gas of baryonic matter falls in through the halo of dark matter it tends to increase its spin rate—just as an ice skater will spin faster when they pull their arms in by their sides. To form a spiral galaxy of about 10 kpc radius (which is a typical size for such objects), the matter must have fallen in from an initial halo of about 100 kpc diameter. In addition, if the quantity of baryonic matter is about one-tenth of the mass of the cold dark matter in the halo, the gravitational effects of the two masses balance to form exactly the smooth sort of rotation curve that we saw in section 14.2.1.

Further details of galaxy formation require complex computer calculations including models of how the hot gas flows into the central region of the dark halo. The process by which the radiative cooling takes place is also significant. (Interestingly this turns out to have been most effective in the early universe when the general density was higher—supporting the idea of early galaxy formation.) However, the general trend seems to be as follows.

1. Central bulges form first as the gas flows into the centre of the dark clump. Stars begin to form as this gas collapses down into smaller pockets.
2. Discs gradually form as more gas flows in round the condensing central bulge.
3. Faint elliptical shaped galaxies can form if two central bulges merge early in their evolution.
4. Large bright elliptical galaxies form from the collisions of fully formed spiral galaxies.

Cold dark matter clumping together gravitationally to forms the seeds of galaxy formation stands as a highly successful model. However, there are problems with this simple idea[16].

Firstly, cold dark matter is very good at forming galaxies, but on its own it is not very good at forming the very large structures that we see in the galaxy distribution. One way out of this difficulty is to invoke the idea of *biasing*.

Cosmologists find it convenient to analyse the density variations in the early universe in terms of 'waves' of density. Think of the surface of a lake in which there is a large amount of swimming, skiing and other mucking about in water going on. A frozen picture of part of the lake would show a very complicated pattern of wave heights. One way of analysing this pattern would be to try and break it down into a set of different 'frozen' waves of different wavelengths. These waves would add up in some parts of the lake and cancel each other out in other places. With a great number of different waves of various wavelengths and amplitudes the result would be a complicated pattern. It can be shown mathematically that any pattern of surface heights can be analysed in this way. Now, these surface heights could represent density variations on a graph of density against position in the universe. Consequently it is also possible to break these density variations down into regular variations of different wavelengths.

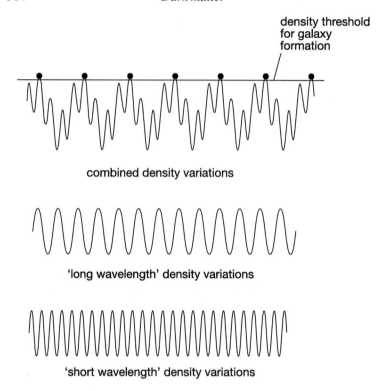

Figure 14.17. Biasing in the formation of galaxies. The density variations can be broken down into 'waves' of various wavelengths (only two components are shown here). Where the peaks of the waves meet, a higher than threshold density fluctuation is formed which is a spot where a galaxy can form.

With these 'waves' of density in the universe it is possible that galaxies will only form where the peaks of the waves combine to form the greatest density concentrations.

Putting this idea into the computer simulations shows that some of the large-scale structure seen in the universe can be reproduced. This would imply that the voids seen are not empty of all forms of matter, but contain huge reservoirs of dark matter. An interesting confirmation of this idea would be the discovery of very faint galaxies in these voids where some baryonic matter collapsed. However, another piece of the puzzle still needs to be filled in before the simulations work to produce all the sorts of structure seen. This is discussed at the end of chapter 15.

After all this, there is still one other outstanding issue that the cold dark matter model does not address—where did the density fluctuations in the early universe come from in the first place?

In this chapter we have concentrated on the issue of dark matter in the universe. Unfortunately, this is not the only problem that faces traditional cosmology. The Friedmann models of the universe have come under threat from two different directions—theoretical and experimental. These issues will be taken up in the next chapter, by the end of which we will be able to put the last piece of the puzzle in place and show how successful the cold dark matter model has been.

14.4 Summary of chapter 14

- Measuring the abundances of various primordial elements strongly suggests that the amount of baryonic matter in the universe adds up to $\Omega_B \sim 0.04$;
- there is a great deal of evidence for the existence of dark matter in the universe (matter that cannot be seen electromagnetically);
- the rotation curves of stars in spiral galaxies can only be explained if the galaxy is embedded in a huge volume of dark matter;
- galactic motion within clusters can only be explained if the cluster is filled with more mass than can be seen visibly;
- gravitational lensing by galaxy clusters reveals the presence of dark matter;
- the redshift of galaxies on the edges of clusters shows that they are being pulled gravitationally into the cluster by a greater amount of mass than can be accounted for visibly;
- dark matter provides a basis for solving the problems of galaxy formation, specifically how did the large-scale structures (filaments and voids) come about and how did galaxies manage to form so early in the universe?
- the COBE map of the microwave background radiation has found very small variations in temperature across the sky;
- the temperature variations reflect differences in the matter density of the universe at the time of recombination;

- dark matter might help explain galaxy formation as its clumpiness would not show up in the COBE map, and so it could be far more clumped than the ordinary matter;
- hot dark matter (neutrinos) does not work as it can only form very large structures;
- cold dark matter works well at forming galaxies early and goes some way to explaining the large-scale structures (especially if biasing is used).

Notes

[1] The name quasar comes from quasi-stellar radio source. Quasars were first discovered from their immensely powerful radio transmissions—greater than those of some galaxies. However, when optical astronomers looked in the places indicated by the radio people, they found objects that looked far more like stars than galaxies. As there was no way that single stars could generate so much power their existence became a mystery. Now we know that the majority of quasars are not emitting radio waves, the name has been changed to quasi-stellar objects (QSOs), but the pronunciation seems to have stuck. An important step came when their light was analysed spectroscopically and found to be unlike anything seen before. Eventually it was realized the light is extremely redshifted ($z > 3$), which is why it took a while to figure out what they are. Consequently quasars must be enormously distant and hence they must have incredibly powerful energy-producing processes inside them to be so brilliant from such distances (typically a thousand times greater than the light output of our galaxy). The current idea is that they are galaxies with enormous black holes at the centre. Recently the observation of supernovae in distant quasars has become an interesting way to measure the expansion rate of the universe.

[2] This is a genuine Doppler shift unlike the Hubble redshift. Astronomers have to subtract the Hubble redshift in order to find the rotation speeds of the stars within a galaxy.

[3] Remember that the parsec, pc, is an astronomical unit of distance equal to about 3.25 light years.

[4] This process is likely to take place within the swirling gas from which the galaxy is formed.

[5] This is the conversion of kinetic energy to gravitational energy that we see when a ball is thrown up into the air.

[6] The estimate of visible matter is made by measuring the light output of the galaxy and comparing it with the light output of a typical star to give the number of stars. Knowing the mass of a typical star gives the mass of the galaxy. The estimation can be refined by including the mass of low intensity stars (so called brown dwarfs) and the dark dust and nebulae in the galaxy. The answer comes up short of that required to explain the rotation curve by a factor of four. This is much too big a gap to be explained by inaccurate estimation.

[7] Of course the supercluster is not 'in' the constellation. The stars that make up Virgo belong to our own galaxy. When an astronomer says an object like a galaxy is in a certain constellation they mean that one has to look out of the galaxy in the direction towards the constellation.

[8] This famous test of general relativity involved taking two pictures of a star field. The first was taken when the Sun was well away from the region of interest and the second when the Sun was close to the star concerned. In order for the star to be visible in the second picture when it was so close to the Sun it had to be taken during a total eclipse.

[9] 2df stands for 'Two Degree Field' system, which is the instrument mounted on the Anglo Australian Telescope (AAT) used in the survey. The 2df instrument is possibly the most complex astronomical instrument devised. It is designed to allow the acquisition of up to 400 simultaneous spectra of objects anywhere within a two degree field on the sky.

[10] The exact manner in which the density fluctuations imprint themselves on the temperature of the background radiation is very complicated. Some more details will be taken up in the next chapter when the origin of the density fluctuations is discussed in more detail.

[11] In an earlier edition of this book I quoted angular resolutions for COBE much smaller than this. Looking back at my research from the time, I can only assume that I had the abilities of two different instruments muxed ip. I apologise for any misunderstanding that this may have caused.

[12] Imagine two points near to each other. One has a large number of neutrinos and the other a smaller number. Random movement between the two points will ensure that the numbers even out over a period of time. Neutrinos are very fast moving objects, so this evening-out process can take place over very large distances. Hence hot dark matter tends to clump together in regions of uniform density that are very large.

[13] Interestingly we can now turn this around. The apparent absence of hot dark matter in the early universe places an upper limit on how great a neutrino's mass can be.

[14] Interestingly the cold dark matter is supported against gravitational collapse by interactions with neutrinos, which exert a pressure—this is a similar mechanism to that which is holding up the baryonic matter, but in that case it is mostly photon pressure.

[15] It is very unlikely that these protogalaxies are all exactly spherical—this gives them 'handles' that the gravitational forces of other protogalaxies can grip onto tending to spin them up as they shrink down in the process of self-gravitational collapse.

[16] In an earlier edition I suggested that one of the problems was that cold dark matter is very slow moving and so it is difficult to explain why some galaxies are observed with very high speeds. However, I am reliably informed that this is not so much of a problem now. Partly this is because the measurements of the galaxy speeds are now producing smaller values, but also the versions of the cold dark matter models that include a cosmological constant (see chapter 15) are also producing higher speed galaxies.

Interlude 3

A brief history of cosmology

Within a year of publishing the general theory, Einstein attempted to apply his new formulation of gravity to the universe as a whole. To do this he imagined the matter of the universe to be smeared out into a uniform density. This is not as bad an approximation as you might think. Although there is some significant clumping of matter on local scales (stars, galaxies, etc) the distribution of clusters of galaxies across the universe appears to be quite uniform on the greatest possible scales[1].

Having studied the complicated equations in some simple cases, Einstein was puzzled because he could not construct a model universe that was static. All the solutions that he tried seemed to describe an unstable universe that would rapidly start to get bigger or smaller. At the time the prevailing belief was that the universe was in a 'steady state' with no overall change taking place (casual observation of the night sky certainly seems to back this idea up). To conform to the experimental evidence of the time, Einstein modified his equations to allow a motionless universe.

It is well established that gravity, as produced by the ordinary matter and energy in the universe, is always attractive. To maintain the universe in a steady state, Einstein needed to guarantee a repulsive effect to balance the pull of gravity. He did this by adding a further term to his equations. The term is simple and characterized by a constant factor, Λ that is referred to as the *cosmological constant*. The value of this constant is not fixed by the theory (indeed it can be zero, which would reduce the modified equations back to those that Einstein had originally published) and so can be picked to provide the required size of effect. Our understanding of the cosmological term has improved since Einstein's original paper on cosmology, in which he proposed this addition to his earlier equations. It certainly follows the general principles that Einstein developed in producing general relativity, but it had a couple of odd features:

- It implies the existence of a form of energy that fills the whole of space (vacuum energy as we would now call it) exerting a repulsive force on all the matter in the universe. The properties of this vacuum energy can be deduced by the requirement of a repulsive force and the form of the mathematical term that would fit into the theory. When Einstein introduced the cosmological term there was no direct experimental evidence for this energy, other than the perceived need to balance the gravity of ordinary matter and energy[2].
- The force produced by this vacuum energy *increases* with distance. On distance scales typical on Earth or in the solar system a cosmological term would produce no observable effects (just as well, as none have been observed) but between galaxies and in the large-scale universe the force would build up substantially. This allows a small value of Λ to be picked that ensures that it does not show up on smaller scales.

By using his modified equations and tuning the size of the cosmological constant, Einstein was able to produce a model universe that was perfectly balanced between the two effects. This was the first formal cosmological model ever published. However, Einstein did not realize that the universe he pictured was unstable—any small variation in the density of matter present (as might be expect by the drift of galaxies etc) would result in the balance being upset. This would be fine if a restoring force was produced that put the universe back into its stable state, but it could be shown that the reverse happened. If the universe became slightly larger due to the density fluctuation, then the gravitational force would get weaker, but the cosmological force would be bigger. This would push the universe to grow, not go back to the original size. A similar argument showed that the universe would collapse given a small reduction in size to start it going.

Shortly after Einstein published his study in cosmology, Wileim de Sitter (a Dutch astronomer) managed to find another solution to the equations that he thought described a static universe.

Like Einstein, de Sitter was convinced that the universe was static, but he felt that the density of matter in the universe was so small, it would have no effect on what was happening. Consequently, de Sitter took everything out but the cosmological term and solved the resulting equations. Einstein was disappointed that another solution had been found—he hoped that his modified equations of general relativity would only allow a single solution that uniquely determined the nature of the universe.

There is an interesting twist to the saga of the de Sitter solution. It was later discovered that the solution actually describes a universe that is expanding rather than being static! It seems strange now that such an 'obvious' mistake could be made, but this is a testament to the difficulty of working with general relativity. Without any matter in the de Sitter universe, there is no clear way of 'seeing' that it is expanding! It was only in 1923 when Eddington and Weyl considered adding a sprinkling of matter to the de Sitter universe that the expansion inherent in the model became clear. The de Sitter solution eventually slid out of favour, but our interest in it will return when we consider the modern ideas of cosmology in chapter 15.

Einstein's disappointment in the discovery of the de Sitter solution was compounded by the publication of two papers in 1922 and 1924[3]. In these papers Alexander Friedmann found a set of general solutions to both the unmodified Einstein equations and those with the cosmological term. Friedmann's work and the equations that he derived now form the basis of modern cosmological techniques. Einstein was convinced that Friedmann's work had to be wrong and even published a short note claiming to have found a mistake. Unfortunately the mistake was his and less than a year later he had to retract the note.

Einstein still maintained that the Friedmann solutions were not physically relevant even though he could not find any error in their derivation. Friedmann died in 1925 (at the very early age of 37) without knowing how significant his papers would become. At the time his work was largely un-noticed as the general feeling was still that the universe was static.

The next significant step in the story involves a professor of astrophysics and priest called Georges Lemaître who published a paper in 1927 on closed universes (it includes a reference to Friedmann's earliest paper). In this paper Lemaître produced the general equations for the evolution of a closed universe containing any amount of matter, radiation and with a cosmological constant. It also contained a discussion of the redshift of light from distant galaxies (something that was not covered in Friedmann's papers) and reference to a value of the Hubble constant—as Hubble had not yet published his work this suggests that he had access to some inside information from Mt Wilson! Lemaître's work did not have much impact straight away as it was written in French and published in a slightly obscure Belgian journal.

In 1929 Hubble published his data on the redshift of distant galaxies. The following year at the meeting of the Royal Astronomical Society in London, the British physicist Eddington got together with de Sitter and decided that neither the original Einstein solution nor the later de Sitter model were tenable after Hubble's observations. Eddington proposed that a search for expanding solutions be started. Lemaître contacted Eddington to make him aware that such a solution had already been found. Subsequently Eddington arranged for Lemaître's paper to be translated into English (it was published in this form in 1931) and acted as quite a publicist for the ideas it contained.

One of the key aspects of Lemaître's paper was a specific model of the universe constructed to comply with the value of Hubble's constant that he was using. In his original work Hubble quite seriously under-estimated the distances involved and produced a value of the Hubble constant of 150 km s^{-1} million light years^{-1} (these days the more accepted value is 20 km s^{-1} million light years^{-1}). In chapter 12 we related the age of the universe to the Hubble constant. The original value that Hubble produced resulted in a very young estimation of about 2 billion years. This was quite a serious embarrassment, as it did not leave enough time for stars to evolve. Lemaître managed to produce a model universe that accounted for the Hubble constant and still gave a universe that was much older than 2 billion years[4].

In 1932 Einstein and de Sitter published a joint paper in which they specified a model mid-way between the open and closed universes that Friedmann had outlined. Their universe did without a cosmological constant and had a flat geometry—consequently it expanded forever.

Heroic work by Walter Baade and Allan Sandage produced papers in 1952 and 1958 which corrected the Hubble constant to something more like its modern accepted value and at this point interest in the Lemaître model became largely historical.

As far as general relativity was concerned, all the theoretical pieces were in place for cosmology by the end of the 1930s, but still progress in the field was slow. The only piece of experimental evidence to back up the big bang was Hubble's law and the physics community was reluctant to accept the idea of an evolving universe based on such scant evidence.

Perhaps the next key phase came in the late 1940s and early 1950s with a series of papers by George Gamow and his collaborators Ralph Alpher and Robert Herman. They were the first people to take a study of the constitution of the early universe seriously. Starting from a universe of hot neutrons (the simplest starting point they could imagine) Gamow hoped that the decay of the neutrons and the subsequent nuclear reactions would build up all the chemical elements. The idea was ultimately not completely successful as the pioneering work by Hans Bethe (1939) and later Burbage(s)[5], Fowler and Hoyle (1957) showed how elements were built up inside stars. However, Gamow is often regarded as the father of the big bang[6] as his work was the first to try and put particle physics (as it was understood at the time) into the early universe. The theoretical understanding his group developed helped form the modern account of nucleosynthesis. Interestingly Gamow was also the first person to predict that there should be some relic radiation left over from the hot big bang.

In 1964 Robert Dicke suggested to a member of his research group, Jim Peebles, that he consider the consequences of a very hot early phase to the universe. Dicke himself was in favour of an oscillating universe model (in which the universe expands from a big bang and then collapses back again triggering another big bang) and felt that there would be evidence of this from the intense thermal radiation produced at each bounce[7]. Peebles took up the suggestion and essentially developed the ideas of nucleosynthesis again (without being aware of the earlier work of the Gamow group). As part of the work he traced the evolution of the thermal radiation through to the present day and predicted that it should have a black body spectrum with a temperature of about 10 K. In February 1965 Peebles gave a seminar at Johns Hopkins University on this work. At this time Penzias and Wilson were struggling with the radio noise that they could not remove from their antenna. A mutual friend of Peebles and Penzias attended the Johns Hopkins seminar and suggested that the two groups get in touch with each other. The result was back-to-back papers on the discovery of the cosmic background in 1965.

This essentially brings us up to the modern day view of standard big bang cosmology. The development of our understanding of nucleosynthesis coupled to the prediction and discovery of the background radiation cemented the big bang as central to our understanding of cosmology. However, as we have seen, it is not without its problems and as our understanding of particle physics has improved, so have our ideas about the very early universe changed, but only in a manner that enhances rather than replaces the work outlined here.

As to the cosmological constant, Einstein later denounced its insertion into general relativity as being 'the greatest blunder of my life'. The view of history, however, may turn out to be rather different. Over the past few years an increasingly convincing body of experimental evidence has been growing that suggests there is some form of vacuum energy exerting a force and accelerating the universe's expansion. This will be one of the main themes of chapter 15.

Notes

[1] Good evidence for the uniformity of the matter distribution in the early universe is provided by the cosmic background radiation. As we shall see in the next chapter this is influenced by the matter density of the early universe and is itself highly uniform.

[2] The astute reader may have picked up here the implication that evidence for the cosmological constant is rather better now than it was in Einstein's day!

[3] In the first paper Friedmann discusses closed universes with and without the cosmological constant. In the later paper he discusses open universes.

[4] The model works as the universe goes through a period of time in which it is 'coasting' with gravity and the cosmological constant in balance.

[5] A husband and wife team.

[6] Ironically, Fred Hoyle coined the term 'big bang' in a radio programme. Hoyle was one of the architects of the 'steady state' theory of cosmology first published in 1948.

[7] Dicke even got a couple of his research group to set up an experiment to look for the background radiation.

Chapter 15

Inflation—a cure for all ills

15.1   Problems with the big bang theory

15.1.1   The flatness problem

As you will recall from chapter 13, the geometry of the universe is determined by the energy density it contains. If this density is greater than a certain critical value the universe is spatially closed, and if it is less than the critical value then it is spatially open. However, if the universe is blessed with a density precisely equal to the critical value, then the geometry is flat (i.e. Euclidean). Currently, our best estimations of the density of the universe, including the contribution from dark matter, suggests that we live in a universe that is 30% of the critical density—in other words one in which $\Omega_{\text{matter}} = 0.3$.

It seems from these estimations that the universe is spatially open (i.e. infinite) and is quite a long way short of the density required to flatten it. However, things are not quite that simple. Although a value of Ω_{matter} of 0.3 seems comfortably far from $\Omega = 1$, it is actually uneasily close. To make sense of this, you have to realize that Ω changes as the universe evolves.

It is easy to see that the actual density of the universe changes with time—there is a constant amount of matter in an increasing volume of space. What is possibly not so obvious is that the critical density required to close the universe will also change as the universe evolves. Even less obvious will be the fact that the two quantities vary with time in different ways.

If the size of the universe is changing according to the scale parameter S, then its volume must be varying like S^3. Consequently, the universe's matter density will scale like S^{-3}.

The critical density is the amount of energy per unit volume required to curve the universe back on itself into a closed 'surface' (if you prefer Newtonian gravitational ideas, then the critical density is the matter required to provide enough gravitational pull to halt the expansion of the universe). One way of visualizing this is to think of the surface of our expanding balloon. When the balloon is small, the curvature of the surface has to be greater in order to close. With a bigger balloon a bigger surface may be needed to cover it, but the surface will be more gently curved (more nearly flat) at every point. A young, small universe requires a very high matter density to close it, but an older larger universe can be closed without being so curved, so the matter density needed is less. The critical density scales in a rather different way to the density:

$$\rho \propto S^{-3}$$
$$\rho_{\text{crit}} \propto \left(\frac{\dot{S}}{S}\right)^2$$

where $\dot{S}$ is the rate at which S is changing at any time. So, the density of the universe and the critical density vary with the age of the universe—but in different ways. Their ratio—density/critical destiny or Ω—is not constant. If density and critical density evolved in the same manner, Ω would be constant for all time. However, in open and closed universes things are rather different—in both cases as time increases the value moves further and further away from $\Omega = 1$. Note that this does not allow the universe to switch from being open to closed as time goes on. If Ω started greater than one, then it will evolve to remain greater than one, and it will always be less than one if that was the initial value. The fate of the universe was determined at the moment of the big bang and cannot change after that. The interesting question is, what was the ratio initially?

As we have seen, it is very difficult to measure the amount of matter in the universe. Evidence is converging on a value of $\Omega_{\text{matter}} = 0.3$ (i.e. the ratio *now* is 0.3, it would have been different in the past) indicating that we are living in an open universe. However, we could be wrong and the

following argument convincingly demonstrates that, in some manner, we
are very wrong!

Because Ω evolves with time, the equations of general relativity allow
us to calculate the value of Ω at any point in the universe's evolution,
given a value at some known time. Consequently we can feed into these
equations the value of 0.3 at an age of about 15 billion years and run
time backwards (in the equations) to see what Ω would have been very
shortly after the big bang. When we do this, a very disturbing answer
emerges. If we project back to 10^{-49} seconds after the big bang, then at
that moment $\Omega - 1 < 10^{-50}$.

This is such a profound result that it needs restating. Imagine we were
to turn the argument round. God has decided that it would be fun to
baffle the cosmologists by arranging for Ω to be 0.3 after fifteen billion years. To do this he has to fix the value of Ω shortly after the big
bang. What values must he choose? Our calculations tell us that God
had better be extremely precise—the universe must be launched with
Ω just a little bit less than one—in fact it has to be somewhere between 0.999999999999999999999999999999999999999999999999999
and   1.000000000000000000000000000000000000000000000000000.
If God were to miss and fall outside this precise range, then Ω would
now be dramatically different from 0.3—in fact after 15 billion years Ω
would probably be vanishingly small and the universe would be utterly
different.

Now, given this degree of precision in the early universe, cosmologists
come to the conclusion that Ω must have been exactly one at the big
bang. This is an argument both from aesthetics and plausibility. If it
is possible for the universe to be launched with any value of Ω, then it
seems incredible that it should be launched so close to one. It is rather
like imagining that God aimed for $\Omega = 1$ and just missed (but came *very*
close). Equally if there is some law of nature that is forcing a particular
value of Ω, it is difficult to see how such a law can be structured to bring
Ω so close to one, without actually being one.

When the full force of this problem came to be appreciated in the
late 1970s and early 1980s, most cosmologists concluded that our
estimations of Ω must be incomplete—if Ω was so close to one in the
past, then the likelihood is that it was exactly one and always shall be

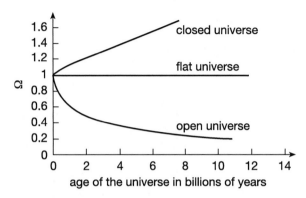

Figure 15.1. This figure shows how Ω changes for open ($\Omega < 1$), closed ($\Omega > 1$) and flat ($\Omega = 1$) universes. The rapid movement away from $\Omega \sim 1$ shows that at the big bang Ω must have been fantastically close to one to be as close as it seems to be now.

(recall that the flat universe is the only case in which the critical density and the actually density evolve in precisely the same manner), so we have another seven-tenths of the universe's energy density still to find.

This argument works just as well for a closed universe. Given a hypothetical value of $\Omega > 1$ now, we can equally calculate what it should have been earlier.

Figure 15.1 shows how Ω changes for open, closed and flat universes. If Ω is exactly one at the initial moment, then it stays exactly one forever and the universe expands in a regular manner. If, however, it is just a bit less than one or just a bit more, then it rapidly diverges as shown. The curves are plotted for a Hubble constant of 23 km s^{-1} per million light years.

Seeing this extraordinary precision in the initial density of the universe, cosmologists suspect one of two things. Either the universe was launched with Ω *exactly* one or there is some physical law at work that drives Ω towards one at some point after the universe starts.

In the community this is known as the *flatness problem*. Why should the universe have been launched with a density so incredibly close to the single unique value that would make it flat for all time?

15.1.2   The horizon problem

If you compare any two points on the sky and make the appropriate
measurements, you will find that the cosmic background radiation
coming from those two points has almost exactly the same equivalent
temperature. As we have seen, this presents a problem for physicists
wishing to understand how galaxies can form—small temperature
variations point to small variations in density, so it is difficult to see how
a region can collapse under its own gravity to make a galaxy. We have
made a great deal of progress in understanding this by introducing the
idea of dark matter. However, these tiny temperature variations present
a problem in their own right. The universe is far too smooth.

To see why this is a problem, imagine that we start with a universe
that is not incredibly smooth and uniform in temperature and set about
establishing this uniformity. Rather fancifully one could think of little
government inspectors intent on making all parts of the universe the
same temperature. These inspectors could have been travelling round
the place since the big bang equipped with heaters and freezers warming
up the cold bits and cooling the hot bits. They have been keeping in
contact with each other by sending pulses of electromagnetic radiation
back and forth. This way they have been able to compare different parts
of the universe and let each other know what the temperature is in their
bit so that they can all come to some agreement.

Perhaps the reason that we see the whole of the sky as being the same
temperature is that they have managed in the 15 billion years the universe
has existed to cover all the parts we can see? In a few million years time
we might start to receive radiation from more distant parts of the universe
that are at a different temperature.

In order to make the activities of our little inspectors correspond to
something that might physically have happened in the early universe,
we need to add one crucial point. Their communication with each other
can only be at best by sending pulses of light back and forth. This
is the fastest way of exchanging information, and they have to work
quickly as the universe is cooling all the time, which makes it harder to
establish the same temperature throughout[1]. Physically, therefore, this
puts a very tight limit on how much time is available for the teams of
inspectors to operate. If they do not complete the job of bringing parts

of the universe to the same temperature before recombination, then they have failed. After recombination there are effectively no free charges left in the universe, so electromagnetic radiation hardly ever interacts with matter any more[2]. This cuts off communication.

Physically what I am suggesting is this. Two objects come to the same temperature when energy flows from one to the other. Energy cannot flow faster than the speed of light, so the best way of exchanging energy is by photons travelling between the objects. However, in the universe at large very few photons interact with matter after recombination. Any mechanism to exchange energy and bring the universe to a uniform temperature had better all be done by recombination, as it will become hopelessly inefficient afterwards.

Using these ideas, we can estimate the size of the region that the inspectors will have managed to bring to the same temperature before they lost contact with each other. Hopefully it will turn out to be at least as big as the universe that we can now see.

The argument runs as follows.

Let us say that the inspectors managed to establish a common temperature right across a region of the universe and the two points at the extreme ends of this region were separated by a distance x_r when recombination happened. That being the case

$$x_r = ct_r \tag{15.1}$$

with t_r being the time between the big bang and recombination. What we want to do is see how big a patch of the sky this region would cover today. Let us assume that the opposite edges of this region were so far away from us at recombination, that the last light they emitted when recombination happened is only just reaching us now. This light must have been travelling towards us for a period of time, t, equal to the age of universe now, t_0, minus the age of universe at recombination, t_r, i.e.

$$t = (t_0 - t_r)$$
$$\sim t_0 \quad \text{as } t_0 \gg t_r$$

(compare 15 billion years with 300 000 years!). Given that the light has taken this long to get to us moving at the same speed, c, for the whole

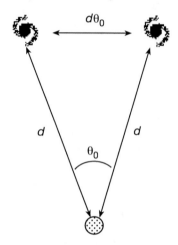

Figure 15.2. Two points on the sky separated by a distance x_0 where $x_0 = d\theta_0$, d being the distance between the two objects and us.

duration, we can estimate that the objects are a distance d away from us now with:

$$d = ct_0. \tag{15.2}$$

Simple geometry allows us to calculate the angle between these objects as seen from Earth right now.

In figure 15.2 I have shown the two points in the sky (drawn as galaxies) and indicated that they are both the same distance, d, from us and a distance x_0 from each other. We observe them to be separated by an angle θ_0 on the sky. If θ_0 is reasonably small (i.e. <1 radian)[3] we can write:

$$x_0 = d\theta_0.$$

The subscript 0 indicating a quantity measured in the current epoch.

Combining this with equation (15.2) gives:

$$x_0 = \theta_0 t_0 c. \tag{15.3}$$

Now we have to link this to x_r—the distance between the objects at recombination. In chapter 12 we found it convenient to scale distances using the parameter S. Generally it is easiest to set $S(t_0) = 1$ (i.e. $S = 1$ now) and so $S(t < t_0) < 1$. Using this we can factor forward from the recombination distance x_r to the separation now x_0:

$$\frac{S(t_0)}{S(t_r)} = \frac{x_0}{x_r}$$

but as $S(t_0) = 1$

$$x_r = Sx_0$$

where I have written $S(t_r)$ as simply S. Bringing equation (15.3) back and putting it in instead of x_0 gives us:

$$x_r = S\theta_0 t_0 c.$$

This result relates the size of the region established by the creatures before recombination, x_r, to the angular size in the sky that this region would represent now, θ_0. We have already established that x_r is limited by the speed of light to being $x_r \leqslant ct_r$ so we can say:

$$S\theta_0 t_0 c \leqslant ct_r$$

$$\therefore \quad \theta_0 \leqslant \frac{t_r}{St_0}.$$

As we are now observing the results of these inspectors' efforts from two points in the sky, we expect them to be at the same temperature provided their angular separation is less than θ_0. Next we must do something to estimate what this angle is. In order to do that we need some way of comparing S at recombination with S now. This we can do by following the temperature variation of the background radiation.

From the previous chapter, the temperature of the universe scales with the parameter S:

$$T \propto 1/S$$

so if the universe was 3000 K at recombination and is about 3 K (the temperature of the background) now, then

$$\theta_0 < \frac{t_r T_r}{t_0 T_0} = \frac{10^5 \text{ yrs} \times 3000 \text{ K}}{10^{10} \text{ yrs} \times 3 \text{ K}} = 10^{-2} \text{ rad} \sim 1°.$$

For comparison it is worth remembering that the full Moon covers an angle of about 0.5° on the sky.

This is a very important result. It shows that any two teams of inspectors that were stationed further apart than one degree in the sky (as we see it now) could not have exchanged information before recombination cut off their means of contact, and so they could not have brought their patches of the universe into the same temperature as each other.

In slightly more physical terms, what we are really saying is that there is no physical mechanism that can bring parts of the sky to the same temperature if they are more than 1° apart, as nothing can travel faster than light and light did not have time to cross this distance before recombination.

Yet the COBE results found patches of the sky separated by much greater angles than this which have the same temperature to better than 0.00001°.

Among the members of the community, this is known as the *horizon problem*—how can two places in the universe be the same temperature if one is 'over the horizon' from the other?

15.2 Inflation

Late on 6 December 1979, Alan Guth (then a post-PhD researcher at SLAC) turned his hand to applying GUT theories to the early universe. The next morning he cycled back into his office clutching his notebook. At the top of one of its pages he scribbled 'spectacular realization'. His calculations, concluded late the previous evening, had shown him a way of solving the horizon and flatness problems in a manner that arose very naturally from particle physics. The grand cooperation between the very large and the very small had paid off in a dramatic way.

Since then Guth's ideas have been refined but the central realization remains.

At the time when the universe cooled to the temperature at which the GUT force broke up into the strong and electroweak forces, the universe must have undergone an extraordinary period of expansion—far faster than anything that had occurred up to that point. So fast, that the universe would double in size every 10^{-34} seconds. Guth termed this rapid expansion *inflation*.

The period of inflation probably started at about 10^{-35} seconds into the history of the universe and stopped at 10^{-32} seconds. However, in this minute period of time the size of the universe would have increased by a factor of 10^{50} or more. That would be equivalent to an object the size of a proton swelling to 10^{19} light years across. If this seems just like any other big number, then consider the following: if the universe is 15 billion years old now, then the part that is currently visible to us is about 15 billion light years across—i.e. 15×10^9 light years—much less than the region that proton-sized dot grew into during inflation!

Let this sink in. I am suggesting that early in its history the universe went through a period during which it expanded so much that part of the universe that before inflation was about the size of a proton ended up bigger than the size of the visible universe now. Of course, this region has still been expanding ever since in the normal manner.

We can turn this around and estimate the size of the currently visible universe at the moment when inflation stopped by taking the scale factor back to that moment. Assuming that the universe is at the critical density after inflation (and as we shall see there are very good reasons for assuming this) we can use the scale parameters of the flat universe (as quoted in chapter 12) to figure out how big it was when inflation stopped.

While matter dominates the universe

$$S \sim t^{2/3}$$

so

$$\frac{S_0}{S_r} = \left(\frac{t_0}{t_r}\right)^{2/3}$$

where 0 indicates the current time, and r indicates the time of recombination. We know that the universe is about 15 billion years old and that recombination happened at about 300 000 years into history,

$$\therefore \qquad \frac{S_0}{S_r} = \left(\frac{15 \text{ billion years}}{300\,000 \text{ years}}\right)^{2/3}$$

$$= 1.36 \times 10^3. \qquad \text{PTO}$$

Using this we can estimate the size of our universe at the time of recombination:

$$\text{Size of universe at recombination} = \frac{15 \text{ billion light years}}{1.36 \times 10^3}$$
$$= 1.11 \times 10^7 \text{ light years.}$$

Now we need to scale this down further from the time of recombination to the time at which inflation stopped (10^{-32} seconds). During this time the universe is radiation dominated, so

$$S \sim t^{1/2}$$

$$\frac{S_r}{S_i} = \left(\frac{t_r}{t_i}\right)^{1/2} = \left(\frac{9.46 \times 10^{12} \text{ seconds}}{10^{-32} \text{ seconds}}\right)^{1/2} = 3.08 \times 10^{22}$$

$\therefore$ size at the end of inflation period is

$$\frac{1.11 \times 10^7 \text{ light years}}{3.08 \times 10^{22}} \sim 3 \text{ m.}$$

One thing is certain, the universe was a lot bigger than 3 m across when inflation ended. Consequently, our visible universe must be a microscopic speck in totality of the universe that is out there (see figure 15.3).

Inflation is such an extraordinary idea it needs to be backed up by some convincing experimental data. It certainly seems to be a consequence of the breakdown of the GUT force, but GUT theory is hardly over burdened with experimental confirmation itself. Results over the past few years are very encouraging, and certainly support the idea of inflation (as we shall see later in this chapter) but the primary reason that inflation became so popular among cosmologists was the very natural way in which it disposed of the horizon and flatness problems. To see this, we need to delve into the physics that drives inflation.

15.2.1 When is a vacuum not a vacuum?

The root cause of the inflationary expansion is the Higgs field introduced by theoreticians to explain how the W and Z particles have mass.

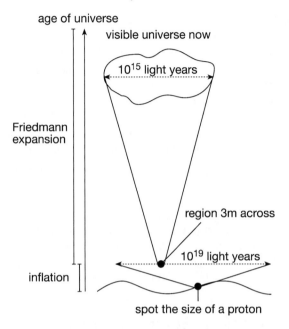

age of universe

visible universe now

10^{15} light years

Friedmann expansion

region 3m across

10^{19} light years

inflation

spot the size of a proton

Figure 15.3. In this hopelessly out of scale diagram I have tried to convey the extraordinary nature of inflation. During inflation regions of the universe the size of a proton expanded until they were 10^{19} light years across. Working back from the whole visible universe now, we can see that this would have been a part of this inflated region that was only 3 m across!

Recall that at high energy the Higgs field cannot be seen, but as the interaction energy drops, so the interaction between Ws, Zs and the Higgs field gives them an effective mass and the perfect symmetry of the electroweak theory is broken. The weak and electromagnetic forces separate.

A similar mechanism is needed to explain how the grand unified force splits into the electroweak and strong forces. In this case it is the X and Y particles that need to be given mass.

GUT theories incorporate a 'super' Higgs field to do this job. At very high energies the perfect GUT symmetry is in operation. Leptons and quarks cannot be distinguished and all the exchange particles are massless. As the interaction energy drops below the point at which the symmetry starts to break (10^{14} GeV), the interaction between the Higgs field and the Xs and Ys gives them a mass of the order of 10^{14} GeV. This breaks the symmetry. The X and Y particles become difficult to emit and the quarks and leptons become distinct. As a consequence the strong and electroweak forces separate out.

There is no hope of ever exploring GUTs directly by experiment. X and Y particles lie well beyond the ability of any conceivable accelerator. However, early in the universe's history the GUT symmetry must have broken.

In the history of the universe, interaction energy corresponds to temperature and temperature varies with time. The energy at which the GUT split into the separate forces must correspond to a moment in the history of the universe. Guth investigated the effect that this change in the status of the super Higgs field would have on the expansion of the universe.

One of the odd things about the Higgs field is that it can still contain a great deal of energy even if its strength is zero. In all field theories the strength of the field at any point is a measure of how likely it is to find an excitation of the field at that point. If the Higgs field has zero strength, then there are no Higgs particles to be found.

It is a little like the surface of a very deep lake. If there are no ripples on the surface, then the lake effectively 'disappears'. However, when a ripple crosses the surface we can see that the water is there. The Higgs field cannot be seen at high energy as no Higgs particles are being produced. However, as the lake has great depth, so the Higgs field can still contain a great deal of energy even though its presence cannot be detected. This energy is latent within the field and cannot be released as there are no Higgs particles.

This energy locked into the Higgs field is the cause of inflation.

The Friedmann equations determine the rate of the universe's expansion by relating it to the amount of energy per unit volume of the universe (the *energy density*). Before GUT theories were introduced cosmologists only included the energy of the matter and radiation present (radiation being defined as relativistically moving objects, i.e. photons, high energy particles, etc). If there was no energy of this form, then the volume of space was a vacuum in the true sense—nothing present (the vacuum perturbations of the fields were not included at this stage either). Solving the Friedmann equations for a matter-dominated universe provided the three basic models discussed in chapter 13. As the universe expands so the energy density decreases (same energy in a bigger volume) and we get a slow expansion with an ultimate fate depending on Ω.

When Guth tackled this problem he realized that the empty vacuum of pre-GUT theory was not empty at all—it contained the super Higgs field. Even though the strength of this field was zero above the GUT temperatures it still contained a great deal of energy and this would have to be included in the energy density of the universe driving the Friedmann equations.

Early in history the energy density of the Higgs field is less than the energy density of the matter and radiation in the universe. The expansion rate is consequently determined by the matter and radiation. However, as the universe expands the energy density of the matter and radiation drops. As the universe reaches the temperature at which the GUT breaks down the Higgs energy density becomes larger than that of the matter and radiation. It takes over controlling the expansion.

This is the moment at which the tremendous period of inflation begins. The Higgs field can drive this rapid expansion as it behaves in a totally different manner to matter or radiation. The energy in the Higgs field does not decrease as the universe expands—it gets bigger!

This marvellous conjuring trick can only really be understood in general relativity. It relies on the fact that gravitational energy is negative. As the universe expands the volume of space increases bringing more Higgs field with it. To create a Higgs field requires energy—positive energy. Conversely as the universe expands the gravitational energy decreases (gets more negative). The total energy remains the same[4]. The one increases at the expense of the other.

To make matters worse, the energy density of the Higgs field is acting like a cosmological constant. In fact, as the Higgs field is dominating over the matter and radiation in the universe, the situation is remarkably

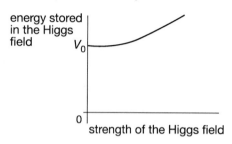

Figure 15.4. The Higgs energy curve at a temperature above the GUT temperature.

similar to that studied by de Sitter earlier this century (see interlude 3). Recall that Einstein introduced the idea of a cosmological constant to balance gravity and produce a static universe. At the time he had no physical model for what the cosmological constant was. Now we do, at least for the early part of the universe's history. The Higgs field is causing the universe to expand at an ever increasing rate as it is exerting an outwards force against gravity. However, at this point the argument closes. The Higgs field is pushing the universe to expand, but as it does so space is getting bigger bringing more Higgs field with it (poaching the necessary energy from the gravitational potential) making the repulsive effect bigger which pushes the universe to expand even more etc. Expansion is running away with the universe.

Once this period of inflation has started the difficult part is stopping it! Once again a remarkable property of the Higgs field comes to the universe's aid.

At temperatures well above the point at which GUTs break down, the energy in the Higgs field varies with its strength in the manner shown in figure 15.4. In order to have minimum energy the Higgs field must be zero everywhere (but note that this minimum energy is not zero).

As the temperature falls so the energy curve of the Higgs field changes to something more like figure 15.5. Now the minimum energy is to have a non-zero Higgs field—in other words to create Higgs particles.

Inflation starts when the energy curve switches from that in figure 15.4 to the one in figure 15.5 and lasts while the field evolves into a state

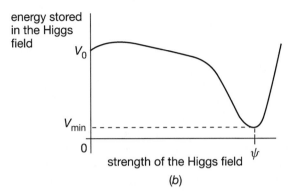

Figure 15.5. (*b*)

energy stored in the Higgs field
V_0
V_{min}
0
strength of the Higgs field ψ

Figure 15.5. The Higgs energy curve at a temperature below the GUT temperature. Inflation is driven by the Higgs field while it is evolving from a value close to zero to something close to ψ, at which point the field breaks down into particles.

of minimum energy. However, as the curve is very gently sloping near to the zero point the change takes place very slowly—far slower than the rate at which the universe is inflating. Consequently, as the inflation is taking place the new space that is being created is forced to contain a Higgs field with an approximately constant energy density. However, eventually the Higgs field gets to a value quite close to the dramatic down curve in figure 15.5 and things start to happen quickly.

The Higgs field suddenly changes to a value ψ shown on the graph and a tremendous amount of energy is released. All the energy that has been put into the Higgs field (at the expense of the gravitational field remember) is now able to escape and form Higgs particles. The symmetry of the GUTs is broken. Now the vacuum is truly a vacuum as the Higgs energy has gone[5]. The inflationary period stops.

In this time the universe has grown by a factor of 10^{50} and it is saturated with Higgs particles of enormous mass. These rapidly decay into more ordinary forms of matter and radiation flooding the inflated universe. The temperature of the universe is increased by this flood of new particles, but not high enough to restore the GUT symmetry.

This is the point at which inflation really works its magic.

We have already seen how inflation takes a tiny speck in the universe and blows it up into a region far bigger than our currently visible universe. The inflation of every tiny subatomic speck is pushed by the Higgs field within that speck. As it grows it creates more Higgs field with the same energy density. Before inflation the Higgs field would have been different in different parts of the universe—it could not be otherwise due to the issue with horizons that I explained earlier. But at subatomic scales the differences would have been extremely tiny. Consequently as the inflation proceeds each speck grows to a universe size with a Higgs field inside it that is virtually identical across the whole of the region. Once that remarkably smooth Higgs field has collapsed, the energy is transferred into Higgs particles and from their decay into 'ordinary' matter. As a result the density of matter is remarkably constant across the whole vast region—exactly what we see in the COBE results. Interactions between the particles set up thermal equilibrium at every point and, because the energy density from the Higgs field is the same across the whole region, the temperature will be the same as well[6]. There is no need for separated parts of the universe to 'communicate' after inflation to set up the same temperature as the matter everywhere has formed from the same field and so has been fed with an energy density that is already virtually identical at every point.

Another remarkable consequence of the inflationary period is that the original matter in the universe has almost totally disappeared. Imagine two quarks in the pre-inflation universe sitting next to each other with about one proton's diameter between them. After inflation they are going to find themselves separated by 10^{19} light years! All the pre-inflation matter has been swept to the farthest corners of the visible universe and beyond. The matter that we can see now (including the stuff we are made of) was not created in the big bang at all—it came from the decaying Higgs field (and so ultimately from gravitational energy).

Here is another piece of inflation magic—the density of the universe before inflation will not have much influence on the universe after inflation. All the pre-inflation matter has been spread out and its density reduced virtually to zero. What now counts is the density of the matter produced by the Higgs field. This was Guth's spectacular realization—that, during inflation, the growth of the universe and the subsequent creation of a new Higgs field would result (after the decay of the Higgs particles) in a universe with a density that was almost exactly the critical density. In other words inflation takes whatever Ω was before and drives it to unity with just the precision needed to solve the flatness problem[7]. One way of seeing why this might be (it really comes straight out of the mathematics) is to think back to our two-dimensional universe on the surface of a balloon. This balloon gets hyper-inflated so the surface at any region looks very much flatter than it did before. As we know, flatness in the universe is determined by the energy density, so the density must be pushed by inflation to be consistent with the curvature as fixed by general relativity—i.e. almost exactly the critical density.

Not a bad night's work all in all.

15.2.2   Ripples in a smooth background

Inflation provides an intriguing explanation for both the horizon and flatness problems, which makes it an interesting theory already. However it does more than that. It also provides an astounding explanation for the density variations in the early universe that eventually give rise to galaxy formation.

As we have seen in earlier chapters, the accepted working model for the formation of galaxies is the gravitational collapse of regions with a higher than normal density. It is clear that the process needs help from dark matter, which can provide pre-existing clumps into which baryonic matter can fall after recombination. However, for the dark matter to clump it must also have density fluctuations built into it and the evidence of COBE strongly points to existing variations in baryonic matter density at the time of recombination. Both these variations need explaining.

For a while it looked as if inflation was not going to help this issue. After all, what inflation does very well is to smooth out the universe and early investigations suggested that it might be too good at this, ironing out any density fluctuations whatsoever. However, further investigation by a variety of different cosmologists suggested that the tiny fluctuations in the Higgs field brought about by quantum processes could lead to just the right sort of density variations.

The quantum nature of any field makes fluctuations inevitable. Consider applying the energy/time uncertainty principle to the Higgs field. The uncertainty in the energy at any point in the field is related to the uncertainty in the length of time that inflation has been taking place in that region of the universe. The exact details of the process are rather complicated as the energy in the Higgs field acts through general relativity to alter the local curvature of spacetime and so helps to determine the duration of the inflationary process. One way of looking at this is to think of the inflation as ending at every point at the same time with fluctuations in the energy of the field, or to think of the process as ending at slightly different times from place to place—which also results in fluctuations.

Generally, quantum fluctuations in a field are confined to a very tiny region and exist over small periods of time. However, during inflation tiny regions are rapidly blown up to enormous sizes. So it is with the quantum fluctuations in the Higgs field. As a fluctuation forms, the region of space it occupies is rapidly blown up to enormous size. The next fluctuation forming in part of the inflating region is also blown up and the process proceeds. Consequently this results in a hierarchy of regions containing Higgs fields with slightly different energies. Inside each fluctuation region lies another and inside that another etc.

During the rapid inflationary period parts of the universe are expanding away from each other at a speed faster than the speed of light. This need not worry us as far as relativity is concerned as no material objects are moving *through* the universe at a speed faster than light. However, because inflation is so fast it does mean that a region smaller than the horizon scale is inflated beyond that so that the ends of the region lose contact with each other.

In our earlier discussion of the horizon problem we touched upon the issues involved in connecting two parts of the universe together. During inflation something rather curious happens. If one discounts any effects due to gravity slowing down the expansion, then it is a reasonable approximation to say that the age of the universe at any time is equal to $1/H$ where H is the Hubble constant at that time. Consequently, as information exchange and physical processes are limited by the speed of light, the horizon scale at any time is c/H—basically any two points in the universe further apart than c/H cannot be linked by any physical process. This is known as the *Hubble horizon*.

Inflation drives the universe to expand at an exponential rate so that:

$$S = \text{constant} \times e^{\Lambda t}$$

and so

$$\dot{S} = \text{constant} \times \Lambda e^{\Lambda t}.$$

(Note that I have, rather suggestively, used Λ in this equation—during inflation the Higgs field is acting exactly like the cosmological constant, Λ, does in the de Sitter universe as it dominates over any matter or radiation density in the universe at the time.)

As Hubble's constant can be expressed as $\dot{S}/S$ (see later for a proof of this) the Hubble horizon during the inflationary period becomes:

$$c/H = c/\Lambda$$

i.e. constant in size.

Figure 15.6 shows the effect this has on one of the inflated regions containing a fluctuation in the Higgs field. Before inflation the universe's expansion was driven by the energy density of radiation (that being the dominant factor—remember that relativistically moving particles can be counted as radiation as well). If we assume a radiation-dominated flat universe then $S \sim t^{1/2}$ and the Hubble horizon $\sim t$. During inflation the Hubble horizon length remains fixed, but a region containing a Higgs fluctuation is expanded from subatomic size to much greater than this horizon length. Once this has happened, no physical process can link opposite ends of the region and so the fluctuation cannot be smoothed out—cosmologists say that the fluctuation has been 'frozen in'. After inflation has ended, this region will be populated by particles produced from Higgs decays and will have a density slightly different from the norm depending on the fluctuation. As the physical size of the region is still very much greater than the Hubble horizon, neither gravity nor the free streaming of particles can smooth out this variation over the region as a whole. It remains a density fluctuation. Of course, once inflation is over, this region will continue to expand with the rest of the universe at a rate dominated once again by the radiation energy density (i.e. $S \sim t^{1/2}$). However the horizon scale is getting bigger at a faster rate (Hubble horizon $\sim t$) so that eventually the region will be completely encompassed by the horizon scale and physical processes can start to modify the density fluctuation.

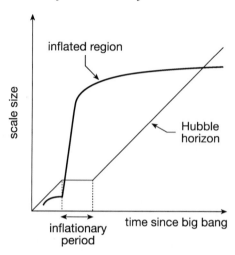

Figure 15.6. During inflation the horizon scale remains the same, while regions of the universe become inflated to much bigger sizes. Once inflation has stopped, the horizon continues to grow at a uniform rate eventually catching up with the inflated region. Consequently a region that has been inflated out of the horizon will eventually re-enter the horizon later in the universe's history.

Not only does inflation provide a mechanism by which variations in density from place to place can be generated and frozen in to the universe for a period of time, it also predicts what sort of spectrum these fluctuations will have. In the previous chapter I discussed how the density variations could be analysed in terms of 'waves'. Basically one can describe the pattern of variations across a region of the universe in terms of a set of waves with 'heights' that determine the density due to that wave. At any point in the universe the density found is the sum of the 'heights' of the waves at that point[8]. If the set of wavelengths of these waves and the amplitude of each wavelength can be found, then the pattern of density variations is completely described.

The best way to follow what happens to the density variations is to deal with the different waves and calculate how their wavelengths are inflated out of the horizon scale, and then what happens to them as they expand normally and the horizon catches up on them. Inflationary theory predicts that the spectrum of the density waves set up by the Hubble field should be 'flat'—i.e. that all the wavelengths should have the same amplitude. This is a prediction that can be tested. Such a pattern of density fluctuations will leave an imprint on the cosmic background radiation[9].

15.2.3 Probing the cosmic background

Undoubtedly the biggest impetus to cosmological research over the next few years will come from the ever finer exploration of the temperature variations in the cosmic background. The ground breaking research of the COBE satellite has already been supplemented by new balloon and ground-based experiments and plans are well in progress for a new generation of satellite measurements.

The cosmic background is an important arena for cosmological research as its fine details allow the experimentalists to test various theoretical models. The density variations that give rise to temperature fluctuations in the background should have the characteristic arrangement predicted by inflation. On a finer level of detail, it is possible to test our understanding of the universe at the time of recombination.

As we have seen, the temperature fluctuations in the background arise due to variations in the density of the material in the universe at the time of recombination, which, in turn, can be traced to quantum fluctuations in the Higgs field that powered inflation. However, this is a broad understanding. The details of how the background radiation map should look at various scales depend on some very complicated process in the super-dense and electrically charged mixture of gas and photons that comprised the universe at the time of recombination.

Broadly the temperature fluctuations can be divided into two types according to how big a region of the sky they cover (not how large a variation in temperature they produce). The super-horizon fluctuations are those which spread over a region of the sky greater than about 1°. Earlier in this chapter I explained how a region of the sky greater than 1° could not have come into thermal equilibrium before the time of recombination as light signals would not have had time to cross a region of that size. (In other words, regions of the universe equal in size to the Hubble horizon at the time of recombination will cover a patch of the sky now about 1° across.) However, inflation would have produced quantum fluctuations on all scales and one would certainly expect to see patches of the sky greater than this angular size showing thermal fluctuations. On sub-horizon scales one should also see fluctuations in temperature, but their detailed structure is more complicated.

[NB: in the following sections when I refer to super- and sub-horizon scales, I am comparing a region's size to that of the horizon scale at the moment of recombination—the Hubble horizon has of course grown much bigger since that time.]

Super-horizon fluctuations

The COBE satellite probed the temperature of the cosmic background across the whole sky but was not able to compare patches smaller than about 7°. This means that the map it produced consisted entirely of super-horizon sized fluctuations (effectively all the finer details were averaged out of the measurements). On these sorts of scales the temperature variations arise from density changes due to the *Sachs–Wolfe effect*.

There are two ways of understanding this depending on how comfortable one is with the ideas of general relativity.

A Newtonian way of looking at the effect would be to realize that regions of higher than average density form local gravitational traps. Photons knocking about in the universe would be pulled into these regions and gain energy in the process (just as a falling ball collects gravitational potential energy and turns it into kinetic energy). Aside from the dark matter, the constituents of the universe leading up to recombination were a super-dense and high-energy combination of protons, neutrons and electrons, so photons would be interacting with the electrons at a very rapid rate and would tend to be held in regions of higher density by these interactions. At the moment of recombination the free electrons get swept up into atoms and the photons can escape from the gravitational traps. However, in the process of escaping they have to expend energy overcoming the gravitational forces holding them back. In between falling in and being released, the universe has expanded, so they have to expend more energy getting out than they gained falling in. Consequently they end up with a net loss of energy. In other words, when these photons are detected eventually by our experiments they appear to have come from a slightly colder patch of the sky. On the COBE map the dark regions correspond to denser parts of the universe at recombination.

The same effect can be viewed from the perspective of general relativity. Regions where the universe is a higher density have a slightly different curvature to those with a lower density. This means that photons that are trapped in such regions (by interacting with the electrons) are not redshifted as much as their neighbours as the universe expands[10]. Effectively they retain more of their original energy than the photons in lower density regions. When they escape at recombination they have to move out of this region of spacetime curvature and so they are influenced by the time curvature on the way out. This has the effect of slowing down their frequencies as detected by us—in other words they lose energy. As before, this is a slightly greater effect, so there is an overall loss of energy from high density regions.

The discovery of the temperature fluctuations by COBE was very important. For the first time we can see direct evidence for early density fluctuations in the universe that will lead (with some aid from the dark matter) to the formation of galaxies. As they are super-horizon sized COBE shows us an image of the conditions in the universe as set down by inflation. There have been no physical processes disturbing the fluctuations that were produced by quantum effects during the inflationary period.

One very important analysis that can be done with the COBE sky map is to look at the angular distribution of the temperature fluctuations. Basically the idea is to look at two patches of the sky separated by some angle θ and calculate the quantity ΔT. The next step is to average this quantity for all pairs of points on the map separated by angle θ. This gives the average fluctuation in temperature at the angular scale θ. Repeating this for all angles from 0 to $180°$ produces a distribution of angular fluctuations such as that shown in figure 15.7.

Note that the *sign* of the temperature variation is not recorded in this plot. A cold spot 30 μK less than average and a hot spot 30 μK greater than average are both recorded as having a ΔT of 30 μK.

In order to make a comparison with the calculations easier, this plot is conventionally done against l (known to its friends as the *multipole*

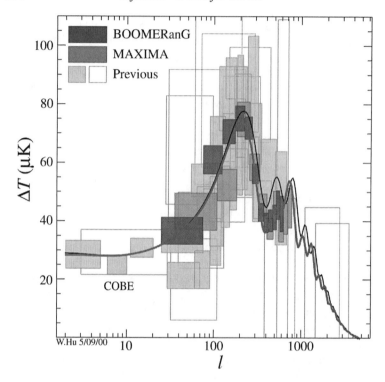

Figure 15.7. A combined plot of various measurements performed on temperature fluctuations of the cosmic background. The horizontal scale can be related to angular sizes using the relationship $\theta \sim 180°/l$. The scale-invariant distribution discovered by COBE dominates the left-hand side of the plot. (Courtesy of Professor Wayne Hu, University of Chicago.)

moment) rather than angle. However the two are easily connected as $l \sim 180°/\theta$. COBE probed the region $2 < l < 20$ and so its measurements dominate the left-hand side of the plot, beautifully illustrating the consequences of a scale-invariant spectrum that has not been disturbed by other processes (as the angles are all greater than those of the horizon at inflation)—the *Sachs–Wolfe plateau*.

COBE's confirmation of the inflationary picture, both through the discovery of temperature fluctuations and that their spectrum agreed with that predicted by inflation, triggered an explosion of experiments probing the background radiation at a variety of different angular scales. The giant peak at $l \sim 200$ is especially important, corresponding to processes going on at sub-horizon scales.

Sub-horizon fluctuations

At the end of the inflationary period the universe contains a super-hot mixture of free charges and photons. In numerical terms the photons outnumber everything else in the universe by a large fraction (if you recall from chapter 12 the ratio of photons to other particles after the antimatter has gone is about 10^9 to 1). This mixture acts rather like a viscous fluid. The photons very easily interact and scatter from the free electrons, and the electrons in turn are attracted to the free positive charges. Consequently all these constituents of the universe are closely coupled to each other and something that disturbs one component also has an influence on the others. The only component that is not strongly coupled to the others is the dark matter that will be lurking about.

The density varies from point to point as a result of quantum fluctuations in a manner most easily described by density 'waves' imprinted on the universe with a variety of different wavelengths. During the post-inflation expansion each wavelength continued to expand in size, but as the Hubble horizon continued to grow at a faster rate they, one by one, became sub-horizon in scale. The first perturbations to form early in the inflationary period grew to enormous sizes and the Hubble horizon did not catch them by the time recombination took place. We now see the effects of these density waves on the COBE map. Fluctuations that formed later did not have as much time to inflate and so they were overtaken by the Hubble horizon before recombination. Consequently various physical processes had time to act on them before recombination cut off the interaction between photons and free charges.

On all scales regions of the universe with a slightly higher than average density tend to pull matter into them from their surroundings and so grow in density. This process is seeded by the existence of clumps of dark matter, which are already beginning to aggregate. (Ordinary matter cannot start to do this significantly until the radiation density has dropped below the matter density—an event that takes place shortly before recombination.) Obviously this tends to increase the gravitational pull such a region exerts. If the region is greater than a certain size (called the *Jeans length*), then the process leads to eventual gravitational collapse when the region has grown in density so that its own self-gravitation becomes irresistible.

On smaller scales the process continues, leading to collapse, but the outcome is different. As the region collapses under gravity, an outward pressure builds up to support the material against collapse. The result is a vibration set up in the material which moves through the universe like a sound wave.

Photons are trapped in regions of high density by the combination of gravitational pull and electron scattering. As with super-horizon-scale regions, these photons have a slightly higher energy than those in surrounding lower density regions. Consequently they exert a net outwards pressure. Furthermore as the region shrinks, the number of collisions between electrons and photons increases, which causes the outward pressure to build up.

The situation is very similar to that found in a sound wave travelling through a gas. A sound wave is comprised of regions of high pressure and density (*compressions*) and regions of low pressure and density (*rarefactions*). During the cycle of a wave, the pressure in the compressions builds up they eventually spring back and expand to form rarefactions. This, in turn, compresses the region next door—compression becomes rarefaction and vice versa. The wave moves through the gas in this sequence.

The scattering of photons by electrons ensures that the two are very strongly linked together, so when the photons push the compressing region outwards, they tend to drag the electrons along as well. The electrons are coupled to the protons by electrostatic attraction, so they also move with the photons.

Consequently, the initial density variation provided by inflation actually triggers 'sound waves' moving through the universe.

Now it is very important not to get these sound waves mixed up with the density 'waves' that I mentioned earlier. The latter are a convenient way of characterizing the density variations in the universe—as such they do not move through the universe. The sound waves triggered by gravitational collapse of small wavelength density variations do travel through the universe at a speed determined by the properties of the photon/charge 'fluid'. However, not all wavelengths of these sound waves survive through to recombination.

The intimate coupling between photons and electrons keeps the two mixed together and forces them to follow the patterns of compression and expansion in one of the sound waves. However, each photon is able to travel a certain distance before it will interact with an electron. This is known as the *mean free path of a photon* (the average distance over which it can travel freely before crashing into something) and it depends on the number of electrons per unit volume in the region. If the wavelength of the sound wave travelling through a region of the universe is about the same as the mean free path of the photons, then that wave will die out. Remember that the sound wave is a pattern of compressed and expanded patches of matter and that a wavelength is the distance between successive compressions. If the mean free path of a photon allows it to travel from one compression to the next without being scattered by an electron, then the photons will easily escape from the compression without providing the necessary outward pressure to turn it into a rarefaction. The wave will die out.

So we now have a picture of what is happening on sub-horizon scales in the period between the matter density becoming the dominant component of the universe and recombination. Sound waves are set up by ordinary matter falling into clumps of dark matter. The first waves to form are on the smallest length scales and many of these are killed off by the flow of photons. The ones that survive may go through several cycles before recombination. Once recombination happens the photons that were trapped in these waves are free to escape and contribute to the cosmic background. However, two processes will have shifted their equivalent temperature. Firstly there is the gravitational/time curvature described earlier that on super-horizon scales leads to the Sachs–Wolfe effect. Secondly the photons will have the imprint of the last scattering process that they experienced. A photon that is scattered off a moving electron will be either red or blue Doppler shifted depending on the direction the electron was moving relative to us observing the scattering. The amount of shift will be governed by the speed.

increasing time from the big bang

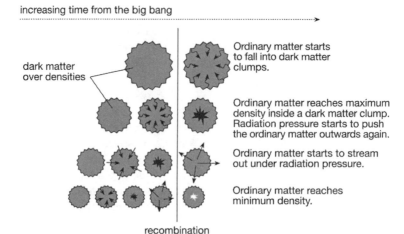

Figure 15.8. Setting up sound waves in the universe before recombination. The process is triggered when ordinary matter is pulled into clumps of dark matter that already exist. As the matter density increases, so the radiation pressure builds up eventually reversing the inward flow and turning a compressed region of matter into a low density region. This has the effect of feeding a nearby compression and setting up a cycle of events. The time scale of this cycle is smallest for the physically smaller length scales—they may go through several cycles leading up to recombination. Note that this diagram suggests that the different size perturbations are separated. In reality they often lie one inside the other.

Consider figure 15.8. On the largest scale (the top row) recombination has caught a region just as the matter was streaming into a dark matter clump. The electron velocity is quite high and so the photons will be released with a large Doppler shift[11]. On the next row (and a smaller scale) we have caught a region just as the matter density reaches its peak and the matter flow is about to reverse due to the radiation pressure. The electron velocities here will be quite small and so there is no Doppler effect; however there will be a large effect due to a Sachs–Wolfe-like contribution. On the next scale down the matter is flowing away from the region and so the photons will be Doppler shifted as they escape. Finally on the last row recombination has happened just as this region has reached its lowest density and once again electron speeds are low, so there is little Doppler shifting of the escaping photons (but a low density Sachs–Wolfe contribution).

Calculating the result of all these effects is complicated and leads to a pattern of temperature variations dominated by a large peak as seen in figure 15.7. The exact multipole value of this peak is influenced by several factors and it is dominated by contributions due to regions like that shown in the second row of figure 15.8. However, it turns out that all the factors due to the size of the region at recombination etc pretty much balance each other and the l value depends quite sensitively on just one factor—Ω. In fact we can say that for Ω values near to 1, $l \approx \frac{200}{\sqrt{\Omega}}$.

The dependence on Ω comes about from the behaviour of light moving through the universe after recombination. All things being equal this peak should be composed of regions about $1°$ across at recombination. However, if the photons have travelled to us from recombination through a curved spacetime this will have the effect of altering the angle. It is rather like looking at the background radiation through a lens. However, for a flat universe there is no dominant curvature of spacetime so we would still expect to see the large peak at $1°$ scales. A brief glance at figure 15.7 shows a dominant peak in the temperature fluctuations at $l \sim 200$ which is very strong evidence for $\Omega = 1$ and a flat universe.

The further peaks provide a rich vein of information that can be mined for detailed measurements of various cosmological parameters. However, the measurements currently available are not sensitive enough to make many useful statements at this stage. There is some evidence in the data for a second peak (caused by catching regions such as the fourth row of figure 15.8 at recombination—remember that in figure 15.7 ΔT represents either an increase or decrease in temperature), but it is possibly not as high as originally expected. Various factors such as the density of baryonic matter in the early universe as well as the amount of dark matter influence the size of the later peaks and investigation of them is likely to become a very hot experimental and theoretical topic over the next few years.

15.2.4 The era of precision cosmology

At the time of writing evidence is still being accumulated from several sources with regard to the temperature fluctuations in the cosmic

−300 μK ▰▰▰▰▰▰▰▰▰▰▰▰▰▰▰▰▰▰▰▰▰▰▱ 300 μK

     300   200   100   0   100   200   300

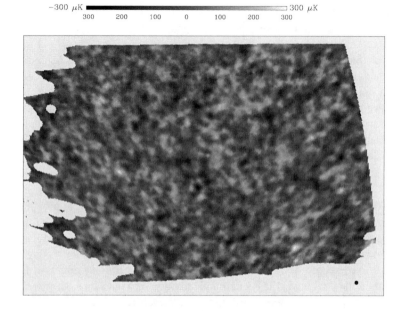

Figure 15.9. This image shows the data produced by the maiden flight of the Boomerang balloon born experiment to measure fluctuations in the cosmic background. Boomerang can only see 2.5% of the sky as measured by COBE, but with 35 times the angular resolution. This data provides good evidence for a peak in the fluctuations as shown in figure 15.7. The small black dot in the lower right-hand corner is the angular size of the full moon for comparison. The image is credited to the Boomerang collaboration and part of it is used on the front cover of this edition.

background. Some of the most impressive data has come from the Boomerang, Max and Maxima experiments, which launched detectors on high altitude balloons (Boomerang was launched in Antarctica and flew for 259 hours, Maxima took off in Texas) carrying instruments to measure the background radiation. There are other ground-based measurements going on a various angular scales and probing different regions of the sky. Plans are also well in hand for more satellite-based experiments.

The Microwave Anisotropy Probe (MAP) was launched on 30 June 2001 and arrived at its specified location 1.5 million kilometres from Earth in October 2001. The hope is that it will be able to map the cosmic background down to angular scales of 0.3° and with a sensitivity of 20 μK. This will be followed up in 2007 by a European Space Agency satellite, Planck, which will aim to make measurements at all angular scales down to at least 0.17°. As a result of the information produced by these experiments we should be able to determine basic cosmological parameters like Ω and H to about 1% accuracy by the end of the decade. This will be an era of precision cosmological measurements brought about by a detailed study of the cosmic background. It will enable some outstanding issues to be settled and should pin down the ultimate fate of our universe. The likelihood is that they will also produce a few surprises.

15.3   The return of Λ

15.3.1   Supernova observations

The observation of the first peak in the cosmic background radiation's temperature fluctuations by the Boomerang and Maxima experiments has provided powerful confirmation of inflation's prediction of a flat universe. However, it does also leave something of a puzzle in that most estimations of the amount of dark matter in the universe place Ω_{matter} at 0.3—there is something else out there making Ω equal to one.

One interesting line of research, that seems to be producing a convincing explanation for this, started to bear fruit at about the same time that the results of Boomerang and Maxima came through—the use of supernovae to chart the history of the universe.

A supernova is a cataclysmic event in which a star explodes. For a brief period of time a star dying in a supernova explosion can outshine the rest of its galaxy put together. This is dramatically demonstrated in figure 15.10.

Supernova explosions can be broadly divided into two classes—type 1 and type 2. Type 2 supernovae take place when old and massive stars can no longer support themselves against their own self-gravitation. The result is a dramatic collapse of the inner core resulting in an explosion that blasts away the outer layers of the star. The remnants of such an explosion can either be a small core comprised entirely of densely

Figure 15.10. This Hubble space telescope image shows a star in the outer spiral arm of a galaxy undergoing a supernova explosion. This is a type 1a supernova and can be seen to rival the galactic core in brightness. The supernova (known as SN1994D) was discovered by the High Z SN search team. (Courtesy the High Z SN search team, the Hubble Space Telescope and NASA.)

packed neutrons (a *neutron star*) or alternatively, if the original star was massive enough, a *black hole*.

Type 1 supernovae happen when a small star (a white dwarf) is gravitationally stealing material from a nearby companion star. Material slowly builds up on the white dwarf until a critical mass limit is reached. When this happens collapse occurs resulting in a supernova explosion. The significant thing about such an explosion is that the critical mass limit is the same in all cases, so that the resulting energy liberation can be assumed to be almost identical in every supernova. Supernovae of this type consequently form *standard candles*, which are the holy grail of cosmological observations. The joy of a standard candle is that they are objects of a known luminosity which can be assumed to be the same everywhere in the universe and for all time. A type 1 supernova going off in a distant galaxy billions of years ago was the same luminosity as a nearby one is now. However, the brightness will be very different, as the light from the distant explosion has had further to travel. This can provide an accurate estimation of how much further away the distant explosion is.

Type 1 supernovae come in several subtypes depending on the details of the light curves that they produce. There are some variations in the peak brightness that they produce, but this can be obtained by observing how the brightness changes with time as the brightest ones fade away more slowly than dimmer ones.

Supernovae are extremely rare objects. The last one to be observed in our galaxy was recorded by Chinese observers in 1054 (the star was so bright it became visible during the day). However, there are an awful lot of galaxies out there and modern observing techniques make it possible to scan a million galaxies in an observing night and find a few tens of supernovae in that time.

Once a supernova has been detected its spectrum is recorded, to provide a redshift, and the brightness measured. If Hubble's law is correct then the brightness (corrected for light being scattered and absorbed by dust on its way to us) and redshift should be directly related. The original aim of these measurements was to look for expected deviations from Hubble's law at high redshift and great distances. What they seem to have found is something rather more surprising. The rate of the universe's expansion is not slowing down, it seems to be accelerating.

15.3.2   Scale and Hubble's law

In order to fully understand the significance of the supernova results it is important to grasp how Hubble's law is an approximation that holds for galaxies that are distant, but still relatively close to us, compared with others[12].

By now the reader should be well used to the idea that the redshift of distant galaxies (as I have been referring to them) comes about due to the stretching of space between the time when the light left the galaxy and when it arrives at our telescopes. It is necessary to use distant galaxies as only then does the redshift due to the Hubble expansion have a bigger effect on the light than the genuine Doppler shift due to the actual motion of the galaxy through space. However, our perspective is now shifting. Figure 12.4 extends the plot of velocity against distance out to 500 MPc showing it to be linear as expected from Hubble's law. However, 500 MPc is only 1.5 billion light years or so. The supernova data extend to 10 billion light years—almost seven times further. Over such distances it is as well to go back to first principles when thinking about redshift.

Imagine that a galaxy, or supernova, emits light of wavelength λ at a time when the scale factor of the universe is S_1. Some time later it is received by us at a longer wavelength λ' when the scale parameter is S_2 (we will fix S_2 to be unity following the normal convention). The new wavelength is related to the old one in the following manner:

If λ was the initial wavelength, then we can say that this is $S_1 \times L$ where L is the length fixed at some reference time. Consequently $L = \lambda/S_1$. Sometime later we receive wavelength λ' which has been scaled up from L so that $\lambda' = S_2 \times L = \lambda \times (S_2/S_1)$. This can be related to the redshift:

$$\text{redshift, } z = \frac{\lambda' - \lambda}{\lambda} = \frac{\frac{S_2}{S_1}\lambda - \lambda}{\lambda} = \left(\frac{S_2}{S_1} - 1\right)$$

is the true equation for the redshift that we must apply. However, if the emission and reception of the light is relatively close in time we can apply an approximation to recover Hubble's law.

$$S_2(t_2) \approx S_1(t_1) + \dot{S} \times (t_2 - t_1)$$

where $\dot{S}$ is the rate at which S is changing with time during that stage of the universe's evolution. Putting this into the redshift equation above gives:

$$z = \left(\frac{S_2}{S_1} - 1\right) = \left(\frac{S_2 - S_1}{S_1}\right) \approx \left(\frac{S_1(t_1) + \dot{S} \times (t_2 - t_1) - S_1}{S_1}\right)$$
$$= \frac{\dot{S} \times (t_2 - t_1)}{S_1}.$$

The distance travelled by the light in this time is $c \times (t_2 - t_1)$, so if we write d for distance we end up with:

$$z \approx \left(\frac{\dot{S}}{S}\right)\frac{d}{c}$$

allowing us to identify Hubble's constant with $\dot{S}/S$.

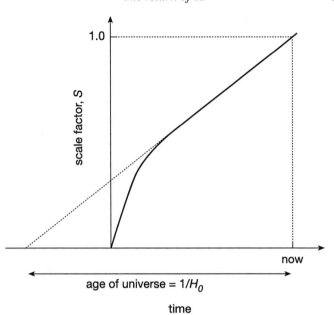

1.0

scale factor, S

now

age of universe = $1/H_0$

time

Figure 15.11. The age of the universe estimated from the Hubble constant now (H_0) is older than the real age if the universe's expansion has been slowing down. The rate of expansion is the gradient on this graph, so the straight region represents a more slowly expanding universe than the curved part.

In chapter 12 I introduced the idea that the age of the universe can be estimated from Hubble's law. Of course it is now clear that this must give us an approximate value, as H has not been constant over the history of the universe.

Figure 15.11 shows how the age of the universe can be overestimated by using $1/H_0$ (H_0 being the value of H at the current time). In this example the scale parameter S is shown to be increasing at a more rapid rate early in the universe's history, as would be expected.

Hubble's constant is $\dot{S}/S$ and with S being set equal to unity in the current epoch, H_0 should be equal to $\dot{S}$. On the diagram, $\dot{S}$ is the gradient of the line at the current time which is the vertical distance to $S = 1$ divided by the horizontal distance marked by the double headed arrow.

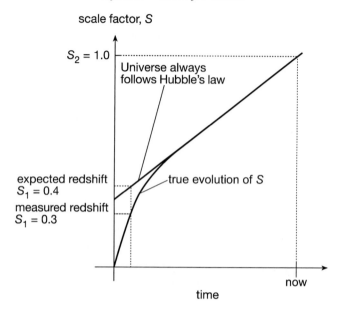

scale factor, S

$S_2 = 1.0$

Universe always
follows Hubble's law

expected redshift
$S_1 = 0.4$

measured redshift
$S_1 = 0.3$

true evolution of S

now

time

Figure 15.12. Assuming that the universe always follows Hubble's law with a fixed value of H leads to the wrong redshift for the most distant objects. The light from these objects left when the scale of the universe was different to that from a linear relationship.

Hence that distance is $1/H_0$ and so the estimated age of the universe. However, as the scale parameter was changing more rapidly with time earlier in the universe's history this turns out to be an over-estimation.

This little analysis has shown that as we probe further and further using more powerful telescopes (and so see back into the past) we should expect deviations from Hubble's law.

Let us take a simple example of this—a flat universe evolving from a big bang. The scale parameter would vary in a similar manner to that shown in figure 15.12.

Projecting back from now with the same value of H would lead us to expect a scale parameter of 0.4 and a redshift of $(S_2/S_1) - 1 = 1.5$. In fact the measured value would be 2.3.

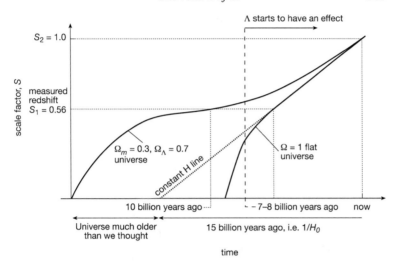

Figure 15.13. In a universe that contains a cosmological constant the expansion will start to accelerate after a certain time. The light from a distant supernova would be dimmer than one might expect given its redshift.

Figure 15.13 shows a more interesting possibility—a universe in which the scale parameter increases but at a reducing rate up to a certain time beyond which it starts to increase at an accelerating rate. Such an effect would follow if the universe (after inflation) contained a small contribution to the energy density from vacuum fluctuations (as described in chapter 11) leading to a cosmological constant. For part of the time the energy density of matter and radiation would dominate and the rate of expansion would be decreasing due to their effect. However, at some time the cosmological constant would start to dominate and consequently the expansion of the universe would commence accelerating.

In this scenario observing a distant supernova would lead to an interesting result. The dimness of the light enables an independent estimation of distance to be made. The redshift comes directly from the spectrum of the light and can be compared to the redshift one would expect from a naive application of Hubble's law. In the case shown in figure 15.13, light was emitted from a supernova when the scale parameter of the universe was about 0.56 (which gives a redshift of

0.8—corresponding to one of the most distant supernovae observed). However, the object appears rather dim compared with what we might expect, leading to a distance measurement of about 10 billion light years. A flat $\Omega_{matter} = 1$ universe would produce a much brighter (i.e. nearer) object for this redshift.

Two competing teams have now published independent data on distant supernovae (the High Z Supernova Search team and the Supernova Cosmology Project). Both report that there is evidence for a cosmological constant that has caused the universe's expansion to start accelerating in the recent past (on a cosmological scale that is!). Although it is still comparatively early days for this research, it seems clear that a non-zero cosmological constant is required in order to explain the data. Figure 15.14 shows a compilation of data from both teams as well as some theoretical models.

The horizontal scale on these graphs is quite straightforwardly the redshift of the light from a supernova (z). The vertical scale will be familiar to astronomers, but not to the less expert reader. It is showing the difference between the *magnitude*, m (which is essentially the brightness of an star, galaxy etc), and the *absolute magnitude* M (which is the brightness the object would have if it were a fixed reference distance away from us). ($m - M$) is therefore a measure of how far away the object really is. Various theoretical curves have been put on the graph for combinations of cosmological parameters.

The lower graph also has z on the horizontal axis, but the vertical axis is now showing the difference between the expected ($m - M$) for a universe with a low matter density (i.e. $\Omega_{matter} = 0.3$), but no cosmological constant. In both graphs the data are indicating that a cosmological constant is influencing the expansion.

If some further research confirms this conclusion (there are always some issues to be worked out as experimentalists gradually understand their results more thoroughly, but most cosmologists are already quite convinced), then theoreticians have to come up with some explanation of this effect. As we discussed in chapter 11, it is not hard to produce a cosmological constant—the vacuum fluctuations of the various quantum fields will do that. The trick is to limit the size of the cosmological constant that they would produce. Initial estimations

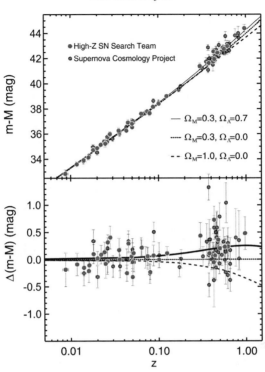

Figure 15.14. The data from distant supernovae. On this graph are the combined measurements of the two main supernova research groups (the High Z Supernova Search team and the Supernova Cosmology Project). The graph and data are credited to both teams.

place the cosmological constant as produced by quantum fields some 120 orders of magnitude greater than the experimental limits. While this is a serious problem, it is far too soon to be throwing away the idea. There is no generally agreed method for calculating the vacuum energy due to vacuum fluctuations, so there is some hope that the various fields will cancel one another to bring the resulting energy density into line. If this is the case, then it does raise a disturbing question similar to one that inflation appears to solve—how can it be that the various quantum fields act to cancel one another with a precision better than one part in 10^{120}? This is another challenge for the future and is probably a pointer to exciting new physics that we have yet to find.

15.3.3 Inflation and Λ

During the inflationary period, the universe was driven by the Higgs field acting as a cosmological constant. At the end of this period the Higgs field disappeared and the universe was dominated by the energy density of the radiation it contained. The effect of inflation should have been to drive the total energy density of the universe towards a value that gives $\Omega = 1$. However, as we have seen it is difficult to account for this in terms of just the matter that we now see the universe containing (including of course the dark matter). The cosmological constant provides a way out of this. If the quantum fluctuations of the fields in the universe provide a vacuum energy that leads to a cosmological constant, this can be fed into the equations governing inflation as well. The size of this effect will be far too small to alter the evolution of the inflationary period, but it will contribute to Ω once it is over. In other words Ω is composed of three parts, Ω_{matter}, $\Omega_{radiation}$ and Ω_Λ such that:

$$\Omega_{matter} + \Omega_{radiation} + \Omega_\Lambda = 1.$$

The radiation component has by now faded away (the background radiation has a very low energy density) but Ω_Λ grows as the universe gets bigger and as the overall geometry of the universe is flat, the sum must equal 1 at all times. Of course, the key question is whether or not inflation is correct and Ω is equal to 1.

We have already seen that the background fluctuations are pointing towards this, but it is interesting to combine the results of those investigations with the supernova data.

When this is done then it seems clear that the two sets of experimental results can be brought into line with each other if the universe has an overall Ω very close to 1 (as would be predicted by inflation) and $\Omega_\Lambda = 0.7$–0.8, $\Omega_{matter} = 0.2$–0.3.

It is now generally agreed that there is sound experimental evidence for the existence of a cosmological constant that is currently causing the expansion of the universe to accelerate.

15.4 The last word on galaxy formation

In the previous chapter I outlined the way in which computer simulations are being used to test the various models of galaxy formation. It is

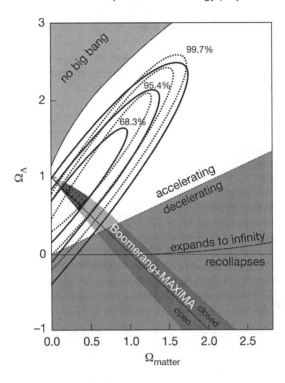

Figure 15.15. A combined plot of the results of measurements of the cosmic background fluctuations (Boomerang and Maxima) with the data from the two supernova teams. The chequered region shows the values of Ω_Λ and Ω_{matter} that are consistent between the two—i.e. $\Omega_\Lambda = 0.7–0.8$ and $\Omega_{matter} = 0.2–0.3$. The data in this graph are credited to all the teams responsible.

clear from the evidence of the cosmic background that a dark matter component is needed to provide some pre-existing structure into which baryonic matter can fall. This is born out by the computer programs which were used to reject hot dark matter, but which are largely successful in explaining the formation of early galaxies using cold dark matter. However, I did point out that there was a factor seemingly missing. The cold dark matter models work well, but do not provide quite the right amount of structure in the clumping to produce the distribution of clusters and super-clusters that we see in the universe. That missing piece is the cosmological constant.

Figure 15.16 shows the results of a set of computer simulations for a variety of different model universes. One can see that the version including cold dark matter and a cosmological constant similar to that indicated by the supernova experiments, has a more open structure. Of course qualitive inspections 'by eye' are not a scientifically valid means of comparison, but the researchers are able to check the degree of structure mathematically and confirm that cold dark matter with a cosmological constant is the most satisfactory model.

15.5   Quantum cosmology

The success of inflation and the use of quantum fluctuations in the Higgs field to explain the density fluctuations has spurred physicists to even greater ambitions. Given that inflation can expand a tiny region of the universe into a vast space, it is natural to wonder if the initial trigger for the universe was a quantum fluctuation from some complex vacuum. Applying quantum theory to gravity brings with it some strange results. The uncertainty principle implies some degree of uncertainty in the geometry of space in a small region. Just as there are amplitudes for various positions and energies, so a quantum theory of gravity contains amplitudes for different spacetime geometries. When one adds to this mix the possibility that the total energy of the universe is zero (balancing the energy of the matter against the negative energy of gravity), it becomes possible that a tiny universe appeared as a fluctuation in some pre-existing quantum reality. Inflation would then expand this proto-universe into the one that we are living in now. This would reduce the whole of our existence to a quantum fluctuation.

This is a very speculative extension of our current theory, but one worth making. It is only by pushing ideas about in this manner that we learn what works and what does not. Perhaps all of the current approaches will turn out to be fruitless, but the exercise is worthwhile for what we can learn from the effort.

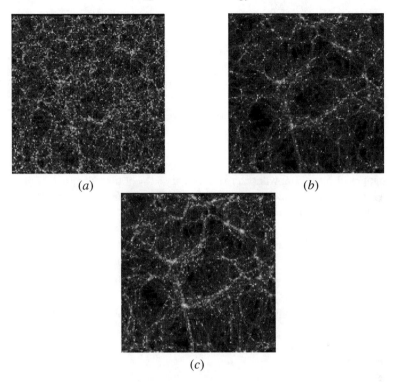

(a) (b)

(c)

Figure 15.16. (*a*) Standard CDM (standard CDM → $\Omega_{matter} = 1.0$, $\Omega_{\Lambda} = 0.0$, $H_0 = 50$ kps/Mpc), (*b*) open CDM (open CDM → $\Omega_{matter} = 0.3$, $\Omega_{\Lambda} = 0.0$, $H_0 = 70$ kps/Mpc), (*c*) lambda CDM (lambda CDM → $\Omega_{matter} = 0.3$, $\Omega_{\Lambda} = 0.7$, $H_0 = 70$ kps/Mpc). Computer graphics produced by a program designed to simulate the evolution of matter under gravity from the early universe up to the present time. The brighter the spot, the denser the clustering of matter is at that point. The figure on the left is generated from a model that contains cold dark matter and with a universe that has a cosmological constant. That on the right is the same, but without the cosmological constant. (These images are courtesy of the 'Virgo consortium for cosmological simulations', a collaboration between the Universities of Durham, Sussex, Cambridge and Edinburgh in the UK, the Max Planck Institute for Astrophysics in Germany and the MacMaster University in Canada. The members of the consortium are H Couchman, G Efstathiou, C S Frenk (Principal Investigator), A Jenkins, J Peacock, F Pearce, V Springel, P Thomas, S White and N Yoshida. The work of the Virgo consortium is carried out on a Cray-T3Es at the Max-Planck Rechen Zentrum in Garching and the Edinburgh Parallel Computing Centre.)

15.6 The last word

Progress in cosmology over the last 20 years has been remarkable. Not so very long ago the subject was not regarded as being the sort of thing that a well brought up physicist got involved with. Worse than a career backwater, it was seen as being dangerously close to philosophy[13]. Once the big bang scenario became well established, all this changed and now cosmology is a thriving and popular field in physics.

Cosmology is becoming such an ordinary branch of science there is a danger of forgetting what an extraordinary subject it is. We are certainly a long way from understanding all the details let alone answering the big questions—such as the following.

Where did the laws of physics come from?

Is it sensible to talk about the universe before the big bang?

Do the laws of quantum theory work in the context of the moment of creation?

What is time?

The last one in particular is a major headache. A fresh understanding of the nature of time would revolutionize much of physics as well as helping us to understand our role in the universe. Admittedly I am prejudiced, but I cannot think of a more exciting subject to be involved with right now.

15.7 Summary of chapter 15

- Standard big bang cosmology has difficulty in explaining how the cosmic background is so uniform in temperature—there was not enough time before recombination for equilibrium to establish itself throughout the universe at the same temperature;
- standard big bang cosmology shows that the critical density must have been extraordinarily close to 1.00 in the very early universe;
- inflationary cosmology provides a neat answer to both these problems;

- during inflation the universe expanded at an extraordinary rate driven by the energy density of a Higgs field;
- inflation caused the Higgs field to have very precisely the same energy density at all points in the inflating universe. Consequently when the Higgs field decayed into matter, equilibrium was established at virtually the same temperature everywhere;
- once the universe had inflated, the density of the original matter had been reduced practically to zero to be replaced by the matter born out of the Higgs field giving a density very close to the critical density (including the energy density of the cosmological constant);
- inflation is also able to explain the ripples in the cosmic background as being due to the inflated quantum fluctuations in the Higgs field;
- studying the cosmic background with high precision has enabled predictions of inflation to be checked. $\Omega = 1$ is indicated by the angular position of the prominent peak in the temperature fluctuations. Higher order peaks when measured properly will enable cosmologists to confirm the details of the effects going on prior to recombination (sound waves imprinting themselves on the temperature fluctuations);
- recent data from the study of distant supernovae confirm that $\Omega = 1$, but point to the existence of a cosmological constant pushing the universe to expand at a faster rate, rather than slowing down as predicted;
- using a cosmological constant in cold dark matter models has helped to explain the creation of structure in the universe.

Notes

[1] By the time I have received the information about the temperature in your part of the universe it has cooled so I am no longer up to date!

[2] Of course, we can still communicate by radio (which moves at the speed of light) as we can make local disturbances in the charges—so stop taking this analogy too literally!

[3] Of course I am slightly anticipating the result here. One radian is about 50°, so these two objects will turn out to be much nearer than opposite sides of the sky!

[4] It is possible that the total energy of the universe is zero. It has always been a worry that the creation of the universe apparently violates the conservation of

energy. However, if the total energy of the universe is zero then the conservation law is obeyed. As the negative gravitational energy increases so the positive Higgs energy increases—still totalling zero. It is the Higgs energy that eventually goes on to form the matter from which we are made. Guth has suggested that the universe might be the ultimate free lunch.

[5] I have not forgotten the vacuum fluctuations! They will enter the story a little later. However, it is easier to simply refer to 'the vacuum' which I will take to include the effects of the quantum fluctuations.

[6] There is a distinction between thermal equilibrium and the same temperature. Each part of the universe could be in equilibrium (within its boundaries), but not necessarily at the same temperature as other parts. The high density of matter produced by the Higgs decays ensures that rapid reactions bring about equilibrium. The constant energy density of the field ensures that the same temperature will result.

[7] This does not mean that it will be *exactly* the critical density. If the universe was greater than critical before inflation, then it will still be greater than critical after inflation: after all inflation simply adds to the pre-existing situation. If it started out to be less than critical then it will still be less than critical after inflation has happened, but very much closer.

[8] It is probably worth reminding the reader of the difference between amplitude and height. The amplitude of a wave is its maximum height. As a water wave travels across the surface of a lake each point on the surface bobs up and down. At any time it has a given height up to a maximum equal to the amplitude.

[9] Picky technical point—scale invariant actually means that as each wavelength is inflated past the horizon scale, it has the same amplitude.

[10] Another way of looking at this, which is basically equivalent, is that from our point of view time seems to be moving differently in the more dense regions. From our point of view, recombination seems to happen sooner in the more dense regions, so the universe has not expanded as much by the time recombination happens. Consequently the photons that are trapped in an over-dense region are not as redshifted as others.

[11] In case you are wondering why we did not bother about the Doppler effect on super-horizon scales, this is because there is no coherent motion of the electron–photon fluid going on. Gravitational pull has not yet got a grip on these regions precisely because they are larger than the horizon scale.

[12] The closest galaxies (e.g. Andromeda) are not always redshifted as their actual movement through space produces a Doppler effect. Light from Andromeda is blueshifted. With more distant galaxies their movement through space is not as significant as the universal expansion—the inherent redshift shows through. In this section we are shifting our horizon again to look at *extremely* distant galaxies.

[13] I have some comments to make about the role of philosophy in physics, and especially cosmology, in the postlude. However to save confusion let me say now that I think the role of philosophy in physics is undervalued by and large.

Postlude

Philosophical thoughts

'The more the universe seems comprehensible, the more it seems pointless'
S Weinberg
'Any universe that contains mind cannot be regarded as pointless'
F Dyson

One thing is clear—frequently when scientists dabble in philosophy they get themselves into trouble. I am probably about to do the same.

Physics is a very beguiling subject. As you work at it or study the works of others it is very difficult not to get caught up in the excitement of the ideas and the conviction that this *must* be right. Ideas have a power—in science just as much as in politics. In science, however, the most powerful ideas also tend to be the most beautiful. Unfortunately the same cannot always be said in politics.

I cannot describe what a beautiful idea is. If you have read this book and been struck by the elegance, simplicity and *rightness* of something, then you already know.

All science aims to describe the universe in which we live—it produces a map of reality. However, when we study an Ordnance Survey map it is not often that we are looking at a picture that seems equally as beautiful as the scenery that we are trying to navigate through. In science the situation is not the same. In our case the scenery (the true laws of the

374

universe if you like) cannot be clearly seen (it's very foggy out there!), so we often get confused between the map and the real thing. At any moment the map is incomplete and contains parts that do not seem to fit quite right. But if it is a really pretty map, then it is very easy to forget this and concentrate on studying the map rather than the world. As many people have pointed out, all our knowledge about the physics of vibrations will not explain why certain pieces of music have such a devastating emotional effect. The map we are currently looking at does not cover that aspect. The mistake is to forget this. The Ordnance Survey people would get in trouble if they decided to leave out some landmarks on the grounds that they were not important, or that the way in which the map was made did not allow them to be covered.

Given this, what are we to make of the apparent observation that some of the maps we have come up with (often the prettiest) seem to work so well? Einstein was very fond of saying that the most incomprehensible fact about the universe was the fact that it was comprehensible. The very fact that we can make a map is quite intriguing.

Some scientists have suggested that the evolutionary pressures of hunting allowed early man to develop a rudimentary mathematical[1] skill that has steadily refined itself into our ability to produce superstring theory. Others regard this as an unreasonable extrapolation of evolutionary biology.

Perhaps the explanation is not important. It is certainly a very striking thing to be able to sit at a desk with a piece of paper and a pencil and doodle away until out comes a series of numbers that actually tell us something about the creation of the universe. Perhaps this human experience itself is the important thing.

It is a wonderful feeling to be able to calculate and discover something new (some scientists describe the event with almost religious terminology). Hot with the enthusiasm of such success scientists are often prone to describe their discoveries in terms that convey great certainty. Sometimes they go into print producing popular books that are titled to sell—again conveying certainty and finality. What is the general person in the street to make of all this?

Some veer into scientism—the conviction that science can explain everything and any question not amenable to scientific analysis is, by definition, a pointless one (or worse a *meaningless* one). To my mind this is to confuse the map with the reality to such an extent that you are confessing to living in the map rather than the reality.

Others decide that science is a worthless exercise and the prognostications of scientists are not to be trusted. Unfortunately whenever the media trawl up some poor expert to express an opinion on something (genetic engineering, BSE etc) this impression is often reinforced. What is such an expert to do? If they say they are sure of something, and are then proven to be wrong the results can be catastrophic. On the other hand, to say that one is not sure is tantamount to saying that you are not the expert and that they should talk to someone else (a neat way of passing the buck). Science is about ordering relative certainty—we are quite sure that the big bang existed, reasonably sure that inflation happened and not at all sure about the quantum events that gave rise to the universe. Sometimes not being sure is the right answer, not because of lack of expertise but because at this stage it is not possible to be sure.

Occasionally the expert can be caught at the wrong moment (especially if this is the moment of discovery). I suspect that the expert on the cosmic background radiation, who was asked for his reaction to the COBE map being presented for the first time, regrets his comment that it was like seeing the face of God.

The conclusion then for the person in the street must be to tread carefully. To respect the discoveries of science, to listen to the opinions of experts and to allow them to be unsure, but never to blindly believe everything that is said especially when an expert ventures out of their field—into philosophy for example.

At any one time the edges of the map are the most interesting parts. As I mentioned earlier it is very foggy out there and the strong temptation is to extend the map into the foggy area (that is the scientific enterprise after all) along the same lines as the clear part of the map. This can be brilliantly successful (science *works*), but it can be terribly misleading. Let us never forget that there are different types of map that work in different situations. Poking about off-map is an activity that is generally regarded as being philosophical, especially if you are rummaging about in a region that is very far away from the map. Scientists are often deeply suspicious of philosophers, principally due to the lack of apparent progress in the subject compared with science. However, this does not always stop them making philosophical statements—often without realizing it. I have quoted two such statements at the start of this section. In both cases the people concerned were quite aware of what they were doing, but one can often read very well known and successful books written by renowned experts that are philosophically meandering to say the least.

The point that I am trying to make is that philosophy is a different activity to science and that when you move from one area into another the same tools of map-making cannot be trusted to work. I was struck recently while explaining the big bang to a class of general studies students when one of them butted in to say that it must be impossible to understand everything about the big bang and still believe in God. I tried to explain that many people see the elegance, simplicity and power of the ideas as being evidence of God's existence as a rational mind behind the workings of the universe, but was not very convincing. What struck me was that this was a classic case of going off-map, but carrying one's pre-cast ideas along on the journey. The quotes at the start of this section illuminate the same effect. Some are predisposed to believe and see the map in those terms, others are inclined not to believe and that view colours their map as well.

My philosophical thoughts on the matter run on two lines. Firstly that the real trick is to stop looking at the map and to hold it up to the world instead. The second is in the form of a question that I keep coming back to—where do the laws of physics come from?

Notes

[1] Ug lob rock at right angle, will hit animal over there so Ug can eat tonight! This kind of reasoning makes one wonder what sort of mathematics would be developed on a world where the dominant life form was vegetarian.

Appendix 1

Nobel Prizes in physics

In the following list of Nobel Prize awards, contributions directly relevant to particle physics and cosmology are in bold type (this is of course a personal judgement).

Year	Name of winner	Citation
1901	Wilhelm Conrad Röntgen	Discovery of x-rays
1902	Hendrick Antoon Lorentz Pieter Zeeman	Effect of magnetic fields on light emitted from atoms
1903	Antoine Henri Becquerel Pierre Curie Marie Sklodowska-Curie	Discovery of radioactivity Research into radioactivity
1904	Lord John William Strutt Rayleigh	Discovery of argon and measurements of densities of gases
1905	Philipp Eduard Anton von Lenard	Work on cathode rays
1906	**Sir Joseph John Thomson**	**Conduction of electricity by gases**
1907	Albert Abraham Michelson	Optical precision instruments and experiments carried out with them
1908	Gabriel Lippmann	For his method of producing colours photographically using interference

1909	Carl Ferdinand Braun Guglielmo Marconi	Development of wireless telegraphy
1910	Johannes Diderik van der Waals	Equations of state for gases and liquids
1911	Wilhelm Wien	Laws governing the radiation of heat
1912	Nils Gustaf Dalen	Invention of automatic regulators for lighthouse and buoy lamps
1913	Heike Kamerlingh Onnes	Properties of matter at low temperatures leading to the discovery of liquid helium
1914	Max von Laue	Discovery of x-ray diffraction by crystals
1915	William Henry Bragg William Lawrence Bragg	Analysis of crystal structure using x-rays
1916		Prize money allocated to the special fund of this prize section
1917	Charles Glover Barkla	Discovery of characteristic x-rays from elements
1918	**Max Planck**	**Discovery of energy quanta**
1919	Johannes Stark	Discovery of Doppler effect in canal rays and the splitting of spectral lines in magnetic fields
1920	Charles-Edouard Guillaume	Services to precision measurements in physics by his discovery of anomalies in nickel steel alloys
1921	Albert Einstein	Services to theoretical physics especially explanation of photoelectric effect
1922	Niels Bohr	Structure of atoms and radiation from them
1923	**Robert Andrew Millikan**	**Measurement of charge on electron and work on photoelectric effect**

1924	Karl Manne Georg Siegbahn	X-ray spectroscopy
1925	James Franck Gustav Hertz	Experimental investigation of energy levels within atoms
1926	Jean Baptiste Perrin	Work on discontinuous nature of matter especially discovery of sedimentation equilibrium
1927	**Arthur Holly Compton** **Charles Thompson Rees Wilson**	**Discovery of Compton effect** **Invention of the cloud chamber**
1928	Owen Willans Richardson	Work on thermionic phenomena
1929	Prince Louis-Victor de Broglie	Discovery of the wave nature of electrons
1930	Sir Chandrasekhara Venkata Raman	Work on scattering of light
1931		Prize money allocated to the special fund of this prize section
1932	**Werner Heisenberg**	**Creation of quantum mechanics**
1933	**Paul Adrien Maurice Dirac** **Erwin Schrödinger**	**Discovery of new productive forms of atomic theory (quantum mechanics)**
1934		Prize money allocated to the special fund of this prize section
1935	**James Chadwick**	**Discovery of the neutron**
1936	**Carl David Anderson** **Victor Franz Hess**	**Discovery of the positron** **Discovery of cosmic rays**
1937	Clinton Joseph Davisson George Paget Thompson	Discovery of electron diffraction by crystals
1938	Enrico Fermi	Discovery of nuclear reactions brought about by slow neutrons (fission)
1939	**Ernest Orlando Lawrence**	**Invention of cyclotron**

1940– 1942		Prize money allocated to the special fund of this prize section
1943	Otto Stern	Discovery of the magnetic moment of the proton
1944	Isidor Isaac Rabi	Resonance recording of magnetic properties of nuclei
1945	Wolfgang Pauli	Discovery of the exclusion principle
1946	Percy Williams Bridgman	Invention of apparatus to produce extremely high pressures and discoveries made with it
1947	Sir Edward Victor Appleton	Investigation of the physics of the upper atmosphere
1948	**Patrick Maynard Stuart Blackett**	**Development of Wilson cloud chamber and discoveries made with it**
1949	**Hideki Yukawa**	**Prediction of mesons**
1950	**Cecil Frank Powell**	**Photographic method for recording particle tracks and discoveries made with it**
1951	Sir John Douglas Cockcroft Ernest Thomas Sinton Walton	Transforming atomic nuclei with artificially accelerated atomic particles
1952	Felix Bloch Edward Mills Purcell	Development of new methods for nuclear magnetic precision measurements and discoveries made with this technique
1953	Frits Zernike	Invention of the phase contrast microscope
1954	Max Born Walter Bothe	Fundamental research in quantum mechanics The coincidence method and discoveries made with it

1955	Willis Eugene Lamb	Discoveries concerning the fine structure of the hydrogen spectrum
	Polykarp Kusch	Precision measurement of the magnetic moment of the electron
1956	John Bardeen Walter Houser Brattain William Shockley	Discovery of transistor effect
1957	**Tsung Dao Lee** **Chen Ning Yang**	**Prediction of parity non-conservation**
1958	**Pavel Aleksejevic Čerenkov** **Il'ja Michajlovic Frank** **Igor' Evan'evic Tamm**	**Discovery and interpretation of Čerenkov effect**
1959	**Owen Chamberlain** **Emilio Gino Segrè**	**Discovery of the antiproton**
1960	**Donald Arthur Glaser**	**Invention of the bubble chamber**
1961	Robert Hofstader	Studies in the electron scattering of atomic nuclei
	Rudolf Ludwig Mössbauer	Research into the resonant absorption of γ-rays
1962	Lev Davidovic Landau	Theory of liquid helium
1963	Maria Goeppert-Mayer J Hans Jensen	Discovery of nuclear shell structure
	Eugene P Wigner	Theory of atomic nucleus
1964	Charles H Townes Nikolai G Basov Alexander M Prochrov	Quantum electronics and masers/lasers
1965	**Richard Feynman** **Julian Schwinger** **Sin-itiro Tomonaga**	**Development of quantum electrodynamics**
1966	Alfred Kastler	Discovery of optical methods for studying Hertzian resonance in atoms
1967	**Hans Albrecht Bethe**	**Contributions to the theory of energy production in stars**

1968	**Luis W Alvarez**	**Discovery of resonance particles and development of bubble chamber techniques**
1969	**Murray Gell-Mann**	**Discoveries concerning the classification of elementary particles**
1970	Hannes Alfvén	Discoveries in magnetohydrodynamics
	Louis Néel	Discoveries in antiferromagnetism and ferrimagnetism
1971	Dennis Gabor	Invention of holography
1972	John Bardeen Leon N Cooper J Robert Schrieffer	Jointly developed theory of superconductivity
1973	Leo Esaki	Discovery of tunnelling in semiconductors
	Ivar Giaever	Discovery of tunnelling in superconductors
	Brian D Josephson	Theory of super-current tunnelling
1974	**Antony Hewish** **Sir Martin Ryle**	**Discovery of pulsars** **Pioneering work in radioastronomy**
1975	Aage Bohr Ben Mottelson James Rainwater	Discovery of the connection between collective motion and particle motion in atomic nuclei
1976	**Burton Richter** **Samuel Chao Chung Ting**	**Independent discovery of J/ψ particle**
1977	Philip Warren Anderson Nevill Francis Mott John Hasbrouck Van Vleck	Theory of magnetic and disordered systems
1978	Peter L Kapitza	Basic inventions and discoveries in low temperature physics
1978	**Arno Penzias** **Robert Woodrow Wilson**	**Discovery of the cosmic microwave background radiation**
1979	**Sheldon Lee Glashow** **Abdus Salam** **Steven Weinberg**	**Unification of electromagnetic and weak forces**

1980	**James Cronin** **Val Fitch**	**Discovery of K^0 CP violation**
1981	Nicolaas Bloembergen Arthur L Schalow Kai M Siegbahn	Contributions to the development of laser spectroscopy Contribution to the development of high resolution electron spectroscopy
1982	Kenneth G Wilson	Theory of critical phenomena in phase transitions
1983	**Subrahmanyan** **Chandrasekhar** **William Fowler**	**Theoretical studies in the structure and evolution of stars** **Theoretical and experimental studies of nucleosynthesis of elements inside stars**
1984	**Carlo Rubbia** **Simon van der Meer**	**Discovery of W and Z particles**
1985	Klaus von Klitzing	Discovery of quantum Hall effect
1986	Ernst Ruska Gerd Binning Heinrich Rohrer	Design of electron microscope Design of the scanning tunnelling microscope
1987	Georg Bednorz K A Müller	Discovery of high temperature superconductivity
1988	**Leon M Lederman** **Melvin Schwartz** **Jack Steinberger**	**Neutrino beam method and demonstration of two kinds of neutrino**
1989	Normal Ramsey Hans Dehmelt Wolfgang Paul	Separated field method of studying atomic transitions Development of ion traps
1990	**Jerome Friedman** **Henry Kendall** **Richard Taylor**	**Pioneer research into deep inelastic scattering**
1991	Pierre-Gilles de Gennes	Mathematics of molecular behaviour in liquids near to solidification
1992	**Georges Charpak**	**Invention of multiwire proportional chamber**

1993	**Russell Hulse** **Joseph Taylor**	**Discovery of a binary pulsar and research into general relativity based on this**
1994	Bertram Brockhouse Clifford Shull	Development of neutron spectroscopy Development of neutron diffraction techniques
1995	**Martin Perl** **Frederick Reines**	**Discovery of tau-lepton** **Discovery of electron-neutrino**
1996	David M Lee Douglas D Osheroff Robert C Richardson	Discovery of superfluid helium-3
1997	Steven Chu Claude Cohen-Tannoudji William D Phillips	Development of methods to cool and trap atoms with laser light
1998	Robert B Laughlin Horst L Stormer Daniel C Tsui	Discovery of a new form of quantum fluid with fractionally charged excitations
1999	**Gerardus 't Hooft** **Martinus J G Veltman**	**Showing that gauge theories (especially that of electroweak force) are renormalizable**
2000	Zhores I Alferov Herbert Kroemer Jack S Kilby	The development of a new type of semiconductor material that is useful in optoelectronics His part in the development of the integrated circuit
2001	Eric A Cornell Wolfgang Ketterle Carl E Wieman	For getting bosonic atoms to form a coherent assembly (like a laser beam is an assembly of photons) and studying the properties of these Bose–Einstein condensates

Appendix 2

Glossary

amplitude
A quantity calculated in quantum mechanics in order to obtain the probability of an event. Amplitudes must be absolute squared to obtained probabilities. They do not add like ordinary numbers, a fact that explains some of the odd effects of quantum mechanics.

annihilation
A reaction in which matter and antimatter combine to produce energy in the form of an exchange particle. The most common form of annihilation reaction is $e^+ + e^- \rightarrow \gamma$ as employed in the LEP accelerator.

antimatter
Every particle of matter has an antimatter partner with the same mass. Other properties, such as electrical charge, baryon number and lepton number are reversed. If matter and antimatter particles meet they annihilate into energy.

atom
Atoms consist of a positively charged nucleus around which electrons move in complicated paths. The nucleus contains protons and neutrons and has the bulk of the mass of the atom. Nuclei are typically a hundred thousand times smaller than atoms. Atoms are the smallest chemically reacting entities.

baryon
A class of particle. Baryons are a subclass of hadrons, and so they

are composed of quarks. All baryons are composed of three quarks. Antibaryons are composed of three antiquarks.

big bang
Current experimental evidence suggests that the universe originated in a period of incredibly high density and temperature called the big bang. Since that time, space has been expanding and the universe has been cooling. The primary evidence for this comes from Hubble's law and the cosmic microwave background radiation.

black body radiation
Any object that is in thermal equilibrium with the electromagnetic radiation it is emitting produces a characteristic spectrum of wavelengths known as black body radiation. The spectrum depends only on the temperature of the object, not on the material it is made from. The cosmic microwave background radiation has a black body spectrum of temperature 2.78 K.

boson
A particle with a spin that is a whole number multiple of $h/2\pi$. All exchange particles are bosons and all 'material' particles are fermions. Hadrons can be bosons if the quarks inside are in the correct alignment for their spins to add to a whole number.

bottom quark
A member of the third generation of quarks. First discovered in 1977 in the upsilon meson (bottom, antibottom).

brown dwarf
A large gaseous object intermediate in size between the planet Jupiter and a small star. As such, it is not big enough to light up nuclear reactions and become a genuine star. Recent observations of brown dwarfs orbiting stars suggest that such objects may be quite common in the universe.

CERN
The European centre for particle physics research on the Swiss/French border just outside Geneva.

charm quark
The charm quark is a member of the second generation. Any hadron containing a charm quark is said to have the property *charm*.

color
Quarks have a fundamental property called color which comes in three types: red, blue and green. Color plays a similar role in the theory of the strong force as electric charge does in the theory of electromagnetism. Antiparticles have anticolor: antired, antiblue and antigreen. All hadrons must be colorless—a baryon must have one quark of each color, a meson has a color, anticolor pair (i.e. blue, antiblue, etc). Leptons do not possess color, and so do not feel the strong force.

conservation law
The universe contains a fixed amount of energy and although it is split between different types (kinetic, potential, etc) the sum total of all the amounts remains the same for all time. This is conservation of energy. Charge is another conserved quantity. The total amount of charge (counting positive charges and negative charges) must remain the same. Other conservation laws are useful in particle physics such as conservation of baryon number and lepton numbers. Physicists are pleased to find conservation laws as they often express fundamental properties of forces.

cosmic microwave background
The relic radiation left over from the big bang. After the era of recombination the photons in the universe were essentially free from interacting with matter. Consequently these photons have remained in the universe undisturbed since that time, aside from a redshift in their wavelengths due to the Hubble expansion. Experiments are now able to chart variations in the effective temperature of this radiation giving clues about non-uniformities in the density of the universe at the time of recombination.

cosmic rays
These are high energy particles that pour down from space. Many of them have their origin in the solar wind produced by the sun (high energy protons). Some come from other parts of our galaxy. Often these particles react with the atoms in the upper atmosphere producing pions. These pions decay into muons on their way down to earth. The majority

of cosmic rays detected near to the surface are muons produced in this way.

cosmological constant
After Einstein had published his theory of gravity, he applied his equations to the universe as a whole. He discovered that he could not produce a stationary solution without adding another term that had not been in the original equations. This term produced a form of repulsive gravity that could be adjusted to keep the universe stationary. The term was characterized by a constant (now known as the cosmological constant), Λ. Einstein later repudiated this alteration after Hubble's results had shown that the universe was expanding. The cosmological constant is finding favour again as the latest measurements using supernova data suggest that the universe's expansion is accelerating.

coulomb
The fundamental unit of electrical charge. In particle physics the coulomb is too large a unit for convenience and the standard unit is set to be the same as the charge on the proton, written as $e = 1.6 \times 10^{-19}$ coulombs.

critical density
The energy density of the universe that gives it a flat (Euclidean) geometry. If the density of the universe is equal to the critical density, then the universe will expand for ever, but at a rate that approaches zero.

cyclotron
A type of particle accelerator. First designed by E O Lawrence. The device relies on the periodic 'kicking' of a charged particle as it moves in a circular path through a potential difference.

dark matter
Matter that does not interact with electromagnetic radiation and so cannot be astronomically observed directly. Dark matter is needed to explain how stars move inside galaxies. It has also been included in theories of galaxy formation. The inflationary theory implies that most of the universe is composed of dark matter. Hot dark matter particles move close to the speed of light. Cold dark matter particles are moving at speeds very much slower than light.

deuterium
An isotope of hydrogen in which there is one proton and one neutron in the nucleus.

down quark
A flavour of quark from the first generation.

eightfold way
The classification of hadrons devised by Murray Gell-Mann based on plotting diagrams of particle families by mass.

electromagnetism
A fundamental force of nature. Static electrical and magnetic forces are actually simple examples of a more general electromagnetic force that acts on charges and currents. In this book electromagnetism is used as a generic name for electrical and magnetic forces.

electron
A fundamental particle and the lightest of the massive particles in the lepton family.

electron-volt (eV)
A unit of energy. One electron-volt is the amount of energy an electron would gain if it were accelerated through a potential difference of one volt.
$1 \, \text{eV} = 1.6 \times 10^{-19} \, \text{J}$
1 keV is a thousand eV
1 MeV is a million eV
1 GeV is a thousand million eV
1 TeV is a million million eV.

electroweak force
In the 1970s it was shown that the weak force could be grouped into a single theory of particle interactions with the electromagnetic force. The resulting theory refers to the electroweak force. In the theory four fundamental fields couple and combine to produce the W^+, W^-, Z^0 and the γ.

elementary particle
A fundamental constituent of nature. Current theory suggests that quarks and leptons (and their antiparticles) are fundamental particles in that they are not composed of combinations of other smaller particles.

Euclidean geometry
In two dimensions this is the geometry experienced on a flat piece of paper. The angle sum of a triangle in this case is always $= 180°$

exchange particle
A particle that is formed from the interaction between two other particles. For example, a photon that passes between two electrons when they interact electromagnetically is an exchange particle. One of the most important successes of particle physics theory in recent times has been the general theory of exchange particles that developed from QED to give us QCD, electroweak unification and the various GUTs.

Fermilab
The American accelerator centre near Chicago. Fermilab has been a major force in particle physics and most recently was home to the discovery of the top quark.

fermion
A particle with a spin that is a fraction of $h/2\pi$.

filaments
Filaments are structures in space composed of many galaxies and clusters of galaxies that seem to be aligned into long thin 'strings' or filaments.

Feynman diagram
Feynman diagrams are often used to represent different aspects of the processes that go on when particles interact via the fundamental forces. They were originally designed as a way of showing how the various mathematical terms in a calculation linked together.

flavour
A whimsical name for the different types of quark and lepton that exist. There are six flavours of quark (up, down, strange, top and bottom) and

six flavours of lepton (electron, electron-neutrino, muon, muon-neutrino, tau and tau-neutrino).

generation
Quarks and leptons can be grouped into generations by the way they react to the weak force. There are three generations of lepton and three generations of quark.

gluon
The exchange particle responsible for the strong force between quarks. Gluons are to the strong field as photons are to the electromagnetic field.

gravitational lens
Einstein was the first to predict that a very strong source of gravity could focus light in a similar manner to an optical lens. Despite Einstein's pessimism several examples of such lenses have been found. Suitably massive objects can be black holes, galaxies and clusters of galaxies. Various optical effects can be produced, depending on how compact the mass is. Smaller objects, such as brown dwarfs can form microlenses. Their effects can be seen in the variation of the brightness of a star that happens to pass behind them (viewed from Earth).

GUT
Grand unified theories attempt to bring three of the four fundamental forces into one all encompassing theory (not gravity). There are several varieties of GUT available, but they all make similar predictions—for example that the proton is not a stable particle and will decay by a process that does not conserve baryon number. So far there has been no experimental confirmation of proton decay.

hadron
A type of elementary particle composed of quarks. The strong force acts on the quarks in such a way as to ensure that quarks can never be found in isolation—they must always be bound into hadrons. Hadrons can be split into two sub-categories—the baryons (and antibaryons) and the mesons.

Higgs particle
The Higgs field is thought to exist and the interaction between this field and the various elementary particles is what gives rise to their mass. If

the field exists then there should be excitations of the field called Higgs particles (as the excitations of the electromagnetic field are photons). As yet there is no direct experimental evidence for the Higgs. The idea of the Higgs field has proven useful in solving some of the problems of the standard model of cosmology—especially in the context of inflationary cosmology.

inflation
A recent theory that modifies the big bang evolution of the universe. The suggestion is that early in the universe's history it grew very rapidly in size (inflated). The mechanisms that drive this inflation would also have the effect of forcing the universe's density towards the critical value. Consequently, we expect the universe's density to be very close to the critical density now if inflation is true. The amount of visible matter in the universe is a lot less than this, so inflation requires dark matter to make up the rest of the density.

ion
An atom from which one or more electrons have been removed. The resulting particle is positively charged. The electrons which have been removed are also sometimes referred to as ions in this context.

kaon
A strange meson. There are four kaons K^+, K^-, K^0 and $\bar{K}^0$.

kelvin scale
The absolute temperature (or kelvin) scale is a measurement of temperature that is not based on any human definition. The centigrade and Fahrenheit scales are based on arbitrary (but sensible) definitions that we have chosen. Absolute temperature is dictated by the properties of gases. The lowest possible temperature is absolute zero (zero kelvin) which corresponds to $-273\,°C$.

Lagrangian
The difference between the kinetic energy of a particle and its potential energy at any point in space and time is the Lagrangian of the particle. Lagrangians are well known in classical mechanics and Feynman extended their use to become the basis of his version of quantum mechanics.

LEP

Large electron positron collider—a very successful accelerator at CERN. LEP operated from November 1989 until November 2000 colliding e^- and e^+ to create Z^0 particles initially; it was then upgraded to produce W^+ and W^- pairs. Just before it was closed down, LEP experiments started to see some evidence for the long awaited Higgs particle.

lepton

A class of elementary particle. Leptons can interact via all the fundamental forces except the strong force. There are six flavours of lepton to match the six flavours of quark.

LHC

Large hadron collider—the replacement accelerator for LEP at CERN. LHC will collide protons at up to 14 TeV.

lifetime

Unstable particles have a lifetime which is the average time taken for particles of that type to decay. There is no way to determine the moment at which an individual particle will decay as the process is inherently random.

Lobechevskian geometry

In two dimensions this is the geometry on the surface of a complex shape similar to that of a riding saddle. The angle sum of a triangle in this case is $< 180°$.

mass shell

Sometimes a particle can be emitted with a mass different than normal. When this happens the particle is said to be 'off mass shell'. This only happens as an intermediate step in an interaction such as those illustrated in Feynman diagrams—the exchange particle is off mass shell. Such particles are sometimes called virtual particles as they cannot be directly observed.

meson

A type of particle composed of a quark and an antiquark bound together. As they contain quarks, mesons are classed as hadrons.

molecule
A structure composed of a group of atoms bound together. Molecules can be quite simple, such as salt (NaCl) or water (H_2O) or very complex containing millions of atoms such as DNA.

muon
A lepton from the second generation. Muons are effectively heavy electrons differing only in that they have muon-number rather than electron-number. Muons are abundant in the cosmic rays that reach ground level.

neutrino
Every massive lepton has an (apparently) massless neutrino with it in a lepton generation. Neutrinos have no electrical charge so they can only interact via the weak force. This makes them very difficult to experiment with as they interact so infrequently. Electron-neutrinos were predicted by Pauli in the 1930s as a way of helping to explain the energy released in nuclear β decay. However, as neutrinos are very weakly interacting, the discovery of the electron-neutrino had to wait until the 1950s when large numbers of them were available from nuclear reactors.

nucleon
General name for any particle contained within a nucleus, i.e. protons and neutrons are nucleons.

nucleus
Part of the atom. The nucleus contains protons and neutrons and so carries a positive electric charge. The electrons of the atom orbit the nucleus in complicated paths. Nuclei are very much smaller than the volume of the atom itself (10^{25} times smaller) yet contain over 99% of the mass of the atom. Nuclei are extremely dense.

Omega (Ω)
The ratio of the actual density of the universe to the critical density. If $\Omega > 1$ then the universe will re-collapse at some time in the future. For $\Omega < 1$ the universe will expand for ever. Inflation predicts that $\Omega = 1$, a prediction that is being born out by recent research into the cosmic microwave background and confirmed by supernova data.

phase
Any quantity that goes through a cyclic motion has a phase which marks the point in the cycle—e.g. the phases of the moon. Phase is normally specified as an angle, with a phase of 2π radians indicating that the cycle is complete. Quantum mechanical amplitudes have both size and phase.

photon
Photons are 'particles' of light. They are the elementary excitations of the electromagnetic field. Photons have zero charge and no mass. The exchange of virtual photons between charged particles gives rise to the electromagnetic force.

pion
The three pions are the lightest particles in the meson class. They are composed of u and d type quarks and antiquarks. Pions are very easily produced in reactions between hadrons.

plasma
A very hot gas in which all the atoms have been ionized.

positron
The antiparticle of the electron.

proton
The lightest particle in the baryon class. Protons are composed of two u quarks and a d quark.

QED, QCD
The theory of the electromagnetic force is known as quantum electrodynamics (QED) and it is one of the most accurate theories that has ever been produced. Its predictions have been experimentally checked to a very high degree of accuracy. Richard Feynman invented Feynman diagrams to make the calculations of QED easier to organize. QED has been used as the model for all theories of the fundamental forces, in particular quantum chromodynamics (QCD) is the theory of the strong force and it was created by making a simple extension to the ideas of QED.

quark
A fundamental particle of nature. Quarks experience all the fundamental

forces. The strong force acts to bind them into composite particles called hadrons. Three quarks make a baryon, three antiquarks make an antibaryon and a quark paired with an antiquark makes a meson. These are the only combinations allowed.

quantum field
A generalization of the wave function and state vector. Quantum fields provide the amplitudes for collections of particles to be created (or annihilated) with their appropriate wave functions. Particles are thought of as 'excitations' of an underlying quantum field.

quantum mechanics
The modern theory of matter that has replaced Newton's mechanics. Quantum theory has many odd consequences including the inability to distinguish particles and having to include all possible paths of a particle when computing an amplitude.

radian (rad)
A measurement of angle based on the circumference of a circle of radius 1: $360°$ is the same as 2π radians, $90°$ is $\pi/2$ radians, etc.

radioactivity
The nuclei of some atoms are unstable. The stability of a nucleus depends critically on the number of protons and neutrons within it. Some isotopes have the wrong number of neutrons for the number of protons. This makes the nucleus unstable. The nucleus can become more stable by getting rid of excess energy in the form of a gamma ray, or by emitting an alpha particle (two protons and two neutrons bound together) or a beta particle (an electron), or a positron. In the latter two cases, either a neutron has decayed into a proton within the nucleus by emitting an electron or a proton has converted into a neutron by emitting a positron.

recombination
Recombination is the name given to the epoch in the universe's history at which electrons combined with the nuclei of hydrogen and helium to form neutral atoms. It is a misnomer in that the electrons had never before been permanently bound to atoms. Before this time the energy in the photons flooding the universe was greater than the energy required to split up atoms, so the electrons were always blasted away. After this

epoch there are essentially no free electrical charges in the universe and photons are much less likely to interact with matter.

Reinmanian geometry
In two dimensions this is the geometry experienced on the surface of a sphere. The angle sum of a triangle in this case is $> 180°$.

renormalization
The mathematical procedure that allows sensible results to be obtained from theories such as QED, QCD and electroweak theory. Some of the Feynman diagrams in such theories 'blow up' giving infinite answers. However, all such diagrams can be swept up by using them to re-define various physical properties (charge, mass magnetic effect), which are then given finite values.

scale factor
If the universe is infinite (as would be the case for $\Omega < 1$), then charting its expansion as a change in size becomes difficult. Cosmologists use the scale parameter, S, to measure the expansion of the universe. Given a pair of galaxies separated by a certain distance, x, now (i.e. at t_0) we can calculate their separation at some time, t, in the future $= S_t \times x/S_0$. As the universe has been expanding, it is conventional to set $S_0 = 1$, so that $S < 1$ for all times in the past.

SLAC
Stanford Linear Accelerator Center—facility in California holding the worlds largest linear accelerator. The SLAC accelerator was used in the pioneering deep inelastic scattering experiments, which revealed direct evidence for the existence of quarks inside protons.

spin
A fundamental property of particles. Spin has no classical equivalent—it is a quantum mechanical property. The closest classical idea is to imagine that particles spin like tops. However, a top can spin at any rate, whereas particles can only spin at a rate that is a multiple of $h/2\pi$. Fermions spin at odd multiples of $h/4\pi$ (e.g. quarks and leptons which spin at $h/4\pi$). Bosons (e.g. photons, gluons, etc) spin at $h/2\pi$ or h/π.

state vector
A collection of amplitudes for a system. The state vector is a similar idea

to that of a wave function. The term state vector tends to be used if there are a finite set of distinct values of a quantity (e.g. energy) involved.

strangeness
Any hadron that contains a strange quark is said to have the property strangeness. This property can be treated a little like lepton number as it is conserved in all reactions that do not involve the weak force. The strangeness of a hadron can be -1 (one strange quark), -2 (two s quarks), -3 (three s quarks), $+1$ (one antistrange quark), 0 (either no strange quarks or an $s\bar{s}$ combination) etc.

strong force
A fundamental force of nature. Quarks feel the strong force as they have the property color. The strong force is mediated by exchange particles called gluons. Gluons also carry color and so are subject to the force themselves. This simple difference between the strong force and the electromagnetic (photons do not have charge and so do not feel the force that they mediate) makes a huge difference in the properties of the force. Whereas the electromagnetic force decreases with distance, the strong force gets stronger. We suspect that because of this the strong force will never allow a quark to be seen in isolation.

superconductors
Materials that have zero resistance to electric current when they are cooled to a low temperature (typically a few kelvin). There is a great deal of research going on at the moment to try to design a material that will be a superconductor at something close to room temperature.

supernovae
These are dramatic stellar explosions caused by a star collapsing under the force of its own gravity. Supernovae can be divided into two types. Type 2 explosions take place when a very massive star collapses under its own weight as its nuclear fuel becomes exhausted. Type 1 supernovae take place when a star has been gravitationally drawing material of a nearby companion star. Once a certain mass limit has been reached, the star feeding off the companion will collapse which triggers a supernova. Type 1 supernovae are interesting cosmologically as the energy released is similar for every explosion. This makes it possible to estimate their distance from the brightness of the light reaching us.

superstring
A hypothetical object that exists in more than four dimensions. Strings are capable of 'vibrating' in various ways to produce the fundamental particles. String theory seems to successfully bring all the fundamental forces together, at the price of formidable mathematical difficulty.

supersymmetry
A theory that relates fermions to bosons. The theory predicts that every fermion particle should have a boson equivalent (e.g. a quark will have a squark) and that every boson should have an equivalent fermion (e.g. photon and photino). There have been some recent hints that supersymmetric particles exist, but as yet none of them have been directly observed.

synchrotron
The first type of particle accelerator was the synchrotron. It relies on the charged particle circulating round in a circular cavity. The cavity is divided into two halves (to form D segments). As the particle crosses from one half to the other it gets a kick from an electric field. This kick can be synchronized (hence the name) to the circulation of the particle, so that it can be repeatedly accelerated.

tau
The heaviest of the lepton family of particles.

thermal equilibrium
A system is in thermal equilibrium when there is no net transfer of energy from one part to another.

top quark
The most recent quark to be discovered. It is very much more massive than any of the others and its discovery took a long time as particle accelerators did not have enough energy to create it.

uncertainty principle
Heisenberg discovered that the amplitudes for position and momentum in a given system were not independent of each other. If a particle were localized in space (so that the amplitudes for it to be found at a point were only significant for a small range of positions) then the amplitudes for the particles momentum would be reasonably large for a wide range of

different momenta. If the momentum were constrained, then the position amplitudes would be spread over a wide range of locations. The same relationship exists for energy and time amplitudes.

unification
Particle physicists hope that at some point in the future they will be able to combine all the theories of the electroweak, strong and gravitational forces into one unified theory. When this happens, the different forces will be seen as different aspects of the one underlying force. There have been some tentative hints that this may be possible and many partially developed theories. Grand unified theories (GUTs) combine the electroweak and strong forces, but we have no experimental evidence as yet to decide which one is right. Theories of everything (TOEs) are more speculative as they include gravity as well. The best candidate TOE at the moment is superstring theory, although this is not fully developed mathematically as yet.

up quark
A flavour of quark from the first generation. Up and down quarks are the most commonly found quarks in nature as they are the components of protons and neutrons.

vacuum fluctuation
A region of space empty of matter and fields is still not a true vacuum. Even if the quantum field has zero intensity in this region, the uncertainty principle tells us that the field still has amplitudes to create particles 'out of the vacuum'. These particles will be created as matter–antimatter pairs and annihilate back into the vacuum. Nevertheless, the existence of these vacuum fluctuations provide a sort of background energy to the universe that may be responsible for a cosmological constant that is pushing the universe into more rapid expansion.

virtual particle
A general name for a particle that is off mass shell.

voids
Gigantic regions in space in which there are far fewer galaxies than in the surrounding space.

W and Z particles
These are the mediators of the weak force. Theories of the weak force contained the W^+ and W^-, but the electroweak theory of Salam, Glashow and Weinberg suggested that the combination of weak and electromagnetic forces required the Z as well if it was to work. The first success of this theory was the discovery of the neutral current—an interaction mediated by the Z.

wave function
Like a state vector, a wave function is a collection of amplitudes for a system (e.g. the position of an electron). However, if the quantity concerned (e.g. position, momentum etc) can have a continuous range of values, then the amplitudes also form a continuous set of values based on that quantity. This is a wave function.

xenon
An element. Xenon is a very unreactive gas and a heavy atom. It has 54 protons in its nucleus. Xenon gas was used in the recent antihydrogen-producing experiment.

Appendix 3

Particle data tables

The quarks

Name	Symbol	Charge (+e)	Mass (GeV/c^2)	Stable	Date	Discovered/ predicted by	How/where
up	u	+2/3	0.33	yes	1964	Gell-Mann Zweig	quark model
down	d	−1/3	~0.33	no	1964	Gell-Mann Zweig	quark model
charm	c	+2/3	1.58	no	1974	Richter *et al* Ting *et al*	SlAC BNL
strange	s	−1/3	0.47	no	1964	Rochester and Butler	strange particles
top	t	+2/3	175	no	1995	CDF collaboration	Fermilab
bottom	b	−1/3	4.58	no	1977	Lederman *et al*	Fermilab

The leptons

Name	Symbol	Charge $(+e)$	Mass (GeV/c^2)	Stable/ lifetime (s)	Date	Discoverers	How/where
electron	e^-	-1	5.31×10^{-4}	yes	1897	J J Thomson	cathode ray tube
electron-neutrino	ν_e	0	$\sim$2–3 eV/c^2	yes	1956	E Cowan F Reines	reactor
muon	μ^-	-1	0.106	2×10^{-6}	1937	S Neddermeyer C Anderson	cosmic rays
muon-neutrino	ν_μ	0	< 170 heV/c^2	yes	1962	M Schwartz *et al*	BNL
tau	τ^-	-1	1.78	3×10^{-13}	1975	M Perl *et al*	SLAC
tau-neutrino	ν_τ	0	< 18.2 MeV/c^2	yes	1978	indirect evidence	—

Some baryons

Name	Symbol	Charge (+e)	Mass (GeV/c^2)	Lifetime (s)	Quarks	Strangeness	Date of discovery
proton	p	+1	0.938	$> 10^{39}$	uud	0	1911
neutron	n	0	0.940	900	udd	0	1932
lambda	Λ	0	1.115	2.6×10^{-10}	uds	−1	1947
sigma plus	Σ^+	+1	1.189	0.8×10^{-10}	uus	−1	1953
sigma minus	Σ^-	−1	1.197	1.5×10^{-10}	dds	−1	1953
sigma zero	Σ^0	0	1.192	6×10^{-20}	uds	−1	1956
xi minus	Ξ^-	−1	1.321	1.6×10^{-10}	dss	−2	1952
xi zero	Ξ^0	0	1.315	3×10^{-10}	uss	−2	1952
omega minus	Ω^-	−1	1.672	0.8×10^{-10}	sss	−3	1964
lamda c	Λ_c	+1	2.28	2.3×10^{-13}	udc	+1 charm	1975

Some mesons

Name	Symbol	Charge (+e)	Mass (GeV/c^2)	Lifetime (s)	Quarks	Strangeness	Date of discovery
pi zero	π^0	0	0.135	0.8×10^{-16}	$u\bar{u}/d\bar{d}$	0	1949
pi plus	π^+	+1	0.140	2.6×10^{-8}	$u\bar{d}$	0	1947
pi minus	π^-	−1	0.140	2.6×10^{-10}	$d\bar{u}$	0	1947
K zero	K^0	0	0.498	*	$d\bar{s}$	+1	1947
K plus	K^+	+1	0.494	1.2×10^{-8}	$u\bar{s}$	+1	1947
K minus	K^-	−1	0.494	1.2×10^{-8}	$s\bar{u}$	−1	1947

* The K^0 and the $\bar{K}^0$ are capable of 'mixing', as the weak force can change one into the other—this makes it impossible to measure their lifetimes separately.

Some more mesons

Name	Symbol	Charge (+e)	Mass (GeV/c^2)	Lifetime (s)	Quarks	Charm	Date of discovery
psi	ψ	0	3.1	10^{-20}	$c\bar{c}$	0	1974
D zero	D^0	0	1.86	4.3×10^{-13}	$c\bar{u}$	+1	1976
D plus	D^+	+1	1.87	9.2×10^{-13}	$c\bar{d}$	+1	1976

Appendix 4

Further reading

There are many excellent books on the market. The following is a list of personal favourites and reference sources used in writing this book.

QED The Strange Theory of Light and Matter
R P Feynman, Penguin Science
ISBN 0-14-012505-1
(Feynman's explanation of quantum mechanics to a lay audience)

The Character of Physical Law
R P Feynman, Penguin Science
ISBN 0-14-017505-9
(This is a book about the laws of nature, but it has a good section on quantum mechanics)

The meaning of quantum mechanics is discussed in:
The Mystery of the Quantum World
E Squires, Adam Hilger
ISBN 0-85274-566-4

Particle and fundamental physics including some relativity is covered in:
To Acknowledge the Wonder
E Squires, Adam Hilger
ISBN 0-85274-798-5

For people who want to push things a little further there are:
Superstrings and the Search for the Theory of Everything
F David Peat, Abacus
ISBN 0-349-10487-5

Gauge Theories in Particle Physics
I J R Aitcheson and A J G Hey, Adam Hilger
ISBN 0-85274-534-6
(The text from which I learned much of my particle physics)

Introduction to Particle Physics
D Griffiths, Wiley
ISBN 0-471-61544-7
(An excellent introductory text that is very readable)

Essential Relativity
W Rindler, Springer
ISBN 0-387-10090-3
(In my opinion the best book on relativity—with an introduction to cosmology)

Relativity: An Introduction to Space-Time Physics
S Adams, Taylor and Francis (a good friend of mine—so read it!)
At press

The Ideas of Particle Physics
J E Dodd, Cambridge University Press
ISBN 0-521-27322-6

Two wonderful books on astronomy, cosmology and the people who work in those fields are:
First Light, the Search for the Edge of the Universe
R Preston, Abacus
ISBN 0-349-104560-5

Lonely Hearts of the Cosmos, the Quest for the Secret of the Universe
D Overbye, Picador
ISBN 0-330-29585-3

Good popular books on cosmology, dark matter and such are:
The Inflationary Universe
A H Guth, Jonathan Cape
ISBN 0-224-04448-6

Wrinkles in Time
G Smoot and K Davidson, Abacus
ISBN 0-349-10602-9

The Shadows of Creation
M Riordan and D Schramm, Oxford University Press
ISBN 0-19-286159-X

The Stuff of the Universe
J Gribbin and M Rees, Penguin Science
ISBN 0-14-024818-8

The Primeval Universe
J V Narlikar, Oxford University Press
ISBN 0-19-289214-2
(More 'meaty' than the others—it includes some maths)

Bubbles, Voids and Bumps in Time: The New Cosmology
J Cornell (ed), Cambridge University Press
ISBN 0-521-42673-1
(Experts in various areas contribute chapters)

Web pages

The following web pages give some of the most up to date information on the work discussed in this book.

Particle physics

CERN home page
http://welcome.cern.ch/welcome/gateway.html
(Where the web was born...)

SLAC virtual visit site
http://www2.slac.stanford.edu/vvc/home.html
(A nice resource for teachers and students)

FERMILAB home page
http://www.fnal.gov/

PPARC particle physics site
http://www.pparc.ac.uk/Ps/Psc/Op/ultimate.asp?Pv=1
(The UK funding body—another useful resource)

Top quark physics
http://www.bodolampe.de/topquark.htm
http://www.sciam.com/0997issue/0997tipton.html

Sudbury Neutrino Observatory (SNO)
http://www.sno.phy.queensu.ca/
(The latest experiment to look for solar neutrinos, producing interesting
data—one to watch)

Super-Kamiokande
http://www-sk.icrr.u-tokyo.ac.jp/doc/sk/
(Japan–US collaboration looking for solar neutrinos)

Cosmology

The 2dF galaxy survey home page
http://www.aao.gov.au/2df/
(Extensive survey on galaxy cluster distributions and redshifts)

Berkeley Cosmology group
http://cfpa.berkeley.edu/
(Good general resource on dark matter etc)

MAP satellite home page
http://map.gsfc.nasa.gov/index.html
(Definitely one to watch)

Boomerang
http://www.physics.ucsb.edu/~boomerang/
(Balloon based experiment to measure CMB fluctuations)

Maxima
http://cfpa.berkeley.edu/group/cmb/index.html
(Another CMB fluctuation experiment)

The MACHO project
http://www.macho.anu.edu.au/
(This is the search for baryonic dark matter in the halo of our own galaxy, which shows up by microlensing stars. The data in figure 14.15 comes from 'Bulge Event 1' listed under the link to Candidate Microlensing Events)

High Z supernova search team
http://cfa-www.harvard.edu/cfa/oir/Research/supernova/HighZ.html

Supernova cosmology project
http://www-supernova.lbl.gov/

Wayne Hu
http://background.uchicago.edu
(Excellent site on CMB physics at all levels)

Index